From Bench to Pilot Plant

ACS SYMPOSIUM SERIES **817**

From Bench to Pilot Plant

Process Research in the Pharmaceutical Industry

Mehdi Nafissi, Editor
Fibrogen

John A. Ragan, Editor
Pfizer Central Research

Keith M. DeVries, Editor
Pfizer Central Research

American Chemical Society, Washington, DC

Library of Congress Cataloging-in-Publication Data

From bench to pilot plant : process research in the pharmaceutical industry / Mehdi Nafissi, editor, John A. Ragan, editor, Keith M. DeVries, editor.

    p. cm.—(ACS symposium series ; 817)

Includes bibliographical references and index.

    ISBN 0–8412–3743–3

    1. Pharmaceutical technology—Congresses. 2. Chemical processes— Congresses.

    I. Title: Process research in the pharmaceutical industry. II. Nafissi, Mehdi, 1936- III. Ragan, John A., 1962- IV. DeVries, Keith M., 1963- V. Series.

    [DNLM: Drug Industry—trends—Congresses. 2. Pharmaceutical Preparations—chemical synthesis—Congresses. 3. Chemistry, Pharmaceutical—methods—Congresses. 4. Drug Design—Congresses. QV 736 F931 2002]

RS192 .F758   2002
615′.19—dc21   2001056743

The paper used in this publication meets the minimum requirements of American National Standard for Information Sciences—Permanence of Paper for Printed Library Materials, ANSI Z39.48–1984.

Foreword

The ACS Symposium Series was first published in 1974 to provide a mechanism for publishing symposia quickly in book form. The purpose of the series is to publish timely, comprehensive books developed from ACS sponsored symposia based on current scientific research. Occasion-ally, books are developed from symposia sponsored by other organizations when the topic is of keen interest to the chemistry audience.

Before agreeing to publish a book, the proposed table of contents is reviewed for appropriate and comprehensive coverage and for interest to the audience. Some papers may be excluded to better focus the book; others may be added to provide comprehensiveness. When appropriate, overview or introductory chapters are added. Drafts of chapters are peer-reviewed prior to final acceptance or rejection, and manuscripts are prepared in camera-ready format.

As a rule, only original research papers and original review papers are included in the volumes. Verbatim reproductions of previously published papers are not accepted.

ACS Books Department

Contents

Indexes

Preface

Achievements of the pharmaceutical industry during the twentieth century at times seem to resemble science fiction, because of advances in related sciences and dedicated scientists who make them happen. Drug discovery is a team effort of scientists from different disciplines, some contributing directly and some indirectly. Physicists' participation, for example, lies in designing instruments that measure molecular parameters that are used for identification and structure confirmation of drug molecules, and quantum mechanical methods help to estimate the energy of interactions between drugs and their biological targets. Biological scientists design in vitro and in vivo models of disease for preliminary screening of candidate chemicals and lead discovery. Pharmacologists study the fate of the drug candidate in living animals. Veterinarians handle animal models in preclinical studies. Pharmaceutical scientists search for a proper formulation that provides maximum absorption as well as study the interaction between the drugs and their packaging materials to prevent degradation. Physicians study the drug in humans to determine efficacy and side effects.

Chemists are involved in all stages of this scientific path: medicinal chemists in the design of the molecule and its synthesis; physical and analytical chemists confirm the structure assignment; computational chemists study structure–activity relationship and participate in the lead optimization; and organic chemists revise the synthetic pathway to make it safe, as well as economically and environmentally suitable for scale-up for extended biological studies and manufacturing.

Most drugs are organic molecules of synthetic origin or natural resources. Prior to the 1970s, many drugs were discovered by random screening of organic molecules. After a lead was discovered it was optimized by structural manipulation to maximize its safety and efficacy.

Recent development of combinatorial and parallel synthetic methods combined with high-throughput screening has opened a new chapter in drug discovery research. Additionally, our knowledge of biochemistry of diseases is expanded to the point that we are able to make molecules that interact mainly with a specific target.

Extended biological studies of this candidate require a substantial amount of the agent. At this stage the scale-up laboratory takes over and an initial quantity of several grams is made that is used for preliminary animal studies by biologists, pharmacologists, and pharmacists.

In the scale-up stage, the synthetic method used is evaluated and modified to meet the cost targets as well as to address environmental and operational safety issues. To achieve these results a process research chemist must optimize the choice and stoichiometry of reagents, solvent, concentrations, temperatures, pH, stir-rates, and many other parameters.

Drugs from natural sources can be classified in two categories: small molecules metabolites and transgenetically produced biopolymers. Small molecule metabolites are extracted from their natural sources by various methods depending on their chemical properties. Often the isolated product is used as a starting material and is modified by chemical manipulations to obtain a desirable biological profile. A good example is 6-aminopenicillanic acid that is used to prepare penicillin derivatives with a better spectrum of activities that alleviate bacterial resistance.

Biopolymers are obtained by genetic manipulations in microorganisms, plants, or animals. This adventure started during the past three decades of the past twentieth century when molecular biologists were able to cleave plasmids, which are circular DNA, at specified positions by a specific enzyme and to insert a strand of selected DNA from other species. A microorganism host is then infected with the modified plasmid. The inserted DNA became part of the host gene and produced the related protein. This method, called recombinant DNA technology, attracted the pharmaceutical industry and as a result the biotechnology industry was formed. The bacterial hosts, however, are not immune to mutation and the inserted gene usually was lost after a few cycles. Domestic animals are targeted for this purpose but are discarded for the fear of cross coupling.

This symposium book provides examples of process development in the pharmaceutical industry by several experts in the field. Chapter 1 is concerned mostly with the general aspect of scale-up such as solvent selection, heating, cooling, and stirring. Chapter 2 is a brilliant example of teamwork. It shows how participation of each member of the team made the project more adaptable to manufacturing. Safety, economy, and environmental impact are always important issues in product development. In Chapter 5, Brummel walks us through different stages of scale-up, expected problems and their projected solutions, and how the development of the multibillion dollar product Lipitor^{TM} was executed.

During the scale-up, involved chemists constantly search for a simpler method that eliminates some steps in the synthetic path. This often results in substantial savings and increases corporate profit. Examples of these are evident throughout the book. Also as we enter the new millennium, with significant advances in the science of organic chemistry, in particular in stereoselective synthesis, manufacturing of racemic drugs are not acceptable any more. Food and Drug Administration encourages evaluation of all stereoisomers and recommends selection of the most active, selective, and less toxic candidates. Chapter 8 is concerned with this aspect of drug development. The book also demonstrates how extensive knowledge of the chemical literature helps a process chemist in solving synthetic problems.

Mehdi Nafissi
Fibrogen Inc.
225 Gateway Boulevard
South San Francisco, CA 94080

Chapter 1

Anatomy of Process Research and Development

Nikolaos H. Petousis

Corporate Development, Ricera, LLC, 7528 Auburn Road,
Concord, OH 44077 (Petousis_N@Ricerca.com)

I.   Introduction

In this report we would like to discuss the importance of process development in the pharmaceutical industry. Superior process development can reduce manufacturing cost, but the power of process development more often lies in how it helps companies achieve a competitive advantage. Consolidation and increased competition in the pharmaceutical industry means that companies are under increased pressure to produce more drugs and to get them to market quicker then ever before. To meet this challenge and gain a competitive edge, most major drug companies are aggressively looking to streamline process development. The benefits of reducing development time are well known. In some cases, short delays in product introduction could be deadly. Perhaps no issue has drawn more attention in recent years than the importance of bringing new products to market faster. Every pharmaceutical manufacturer is only too aware of the rising cost to bringing a drug to market. Indeed, the process of getting a novel drug to market is now estimated to cost the manufacturer more than $500 million from the point at which a candidate is identified[1].

A patent provides a guarantee that a single company will have a short-term monopoly over the production of a new drug. However, competition amongst therapeutic areas as well as "hot" areas such as HIV, pain, diabetes, results in very short-term monopoly. Development folks are under extreme pressure to avoid any delays in getting a drug to market. For example, G. D. Searle's Celebrex was on the market for two months before Merck introduced Vioxx.

In 1999, R&D spending increased by 14 percent[2]. With R&D spending increasing faster than sales, and patent expirations outpacing new chemical entity approvals, the industry is under tremendous pressure to improve the product development cycle. Complex products and chemistries such as potent drugs result in high R&D expenditure. For example, potent drugs require total containment facilities. Such facilities are expensive to build and operate. On the other hand, the worldwide sales of single-isomer chiral drugs have increased from 1 percent in 1985 to 12 percent in 2000[3]. Single isomer chiral drug sales should reach $500 million in 2000[3]. Pharmaceutical sales were estimated at $97 billion in 1998, 11% higher than 1997's $87 billion[4]. The situation intensifies with the demand for lower costs from governments and managed care organizations, generics and competition from newly industrialized countries such as China and India.

The benchmark of success in the pharmaceutical industry being the release of a blockbuster drug, their production now lies exclusively in the hands of a small number of large drug companies that have successfully merged, and consequently seek lean and portfolio specific developments with high levels of return. The increasing level of investment required to maintain an innovative drug company and deliver a significant number of important drugs to the marketplace has illustrated the importance of process research and development in a successful product development cycle.

II. The Drug Development Cycle

Preclinical development:
During preclinical development a company evaluates the identified candidate and the process research group supplies material to the formulators, the pharmacologists and the pharmacokinetics group. Some companies invest in process development, more aggressively than others.

In this stage, several animal models are used to evaluate toxicity and pharmacokinetic properties of the compound. After completing preclinical testing, the company files an Investigational New Drug Application (IND) with the Food and Drug Administration (FDA) to begin to test the drug in humans. The IND becomes effective if the FDA does not disapprove it within 30 days. The IND shows results of previous experiments, how, where and by whom the new studies will be conducted; the chemical structure of the compound; how it is thought to work in the body; any toxic effects found in the animal studies; and how the compound is manufactured. In addition, the IND must be reviewed and approved by the Institutional Review Board where studies will be conducted, and progress reports on clinical trials must be submitted at least annually to the FDA.

The majority of time and resources in the drug development cycle are consumed by the process of evaluating safety and efficacy in human patients. Clinical trials are divided into three phases.

Phase I Trials:
During phase I trials, the drug is administered at several dosages to healthy volunteers to determine whether there are any intolerable side effects. Although even minor side effects, such as a rash at the site of injection, are investigated, they are usually not enough to halt trials. The drug is typically administered in a prototype formulation such as a capsule or injection. These tests take about a year and involve about 20 to 80 healthy volunteers. The tests study a drug's safety profile, including the safe dosage range. The studies also determine how a drug is absorbed, distributed, metabolized and excreted, and the duration of action.

Phase II Trials:
During phase II trials, emphasis shifts to demonstrating the drug's efficacy in controlled studies. In this phase, controlled studies of approximately 100 to 300 patients (people with the disease) assess the drug's effectiveness and take about two years. These studies are typically carried out by giving the drug to one group of patients and a placebo or alternative treatment to a control group. These drugs are often double blind: neither the people administering the drug nor the patients know whether they are receiving the drug, the placebo, or an alternative. It is during this phase that the formulation group designs a number of dosage forms and attempts to determine which performs best in human patients.
During phase II trials, process development has to provide material to two groups; the formulation group, and the clinical development group. The formulation group needs high-quality product for their analytical experiments, while the clinical group needs material for human testing. The demand for the bulk active intensifies as the formulation studies intensify and the number of patients increases.

Phase III Trials:
This phase last about three years and usually involves 1,000 to 3,000 patients in clinics and hospitals. The start of phase III trials is perhaps the most important milestone in the drug development process. The purpose of this phase is to generate definitive data on the safety and efficacy of the drug in large-scale, long

term, multi-site trials. By the time a drug has reached phase III there is an 85 percent probability that it will reach the market.

Phase III clinical trial represents an important milestone to chemical development. During this phase, process development focuses on finalizing all details of the process and proceeding to validation. Data from these validation run, usually three, are submitted to the FDA (Food and Drug Administration) as part of the New Drug Application (NDA) filling.

III.   Scale-up of Chemical Processes

Scale-up is inherent in all industrial activity. Scale-up is constantly on the minds of chemists and engineers concerned with the development of new processes or expansion of existing ones.

When a new chemical process or a change of a process moves from the laboratory to the pilot plant to the manufacturing facility, unexpected problems are often encountered. The problems could be of either chemical or physical nature, or a variation of both. One of the most frustrating difficulties that can be encountered is the presence of impurities that are not considered or studied in the small laboratory scale.

Since scale-up is the theme of this report we have defined it as:

The successful startup and operation of a commercial process whose design is in part based upon experimentation and demonstration at a pilot scale of operation

The concept of successful includes production of the product at planned rates, at the projected cost, and to the desired quality standards. Obviously to meet the desired cost we must consider the purchase prices of raw materials, the product yield, and the return on capital, but also the overall safety of the operation, the personnel, the public and the environment.

To be successful at the scale-up of chemical processes requires the utilization of a broad spectrum of technical skills and a mature understanding of the total problem under study. The chemist's knowledge of the chemical literature and of the raw material database may restrict the choice of routes made. I cannot stress too much the importance of keeping up with the chemical literature as an essential part of the discipline of chemical development. The literature should be surveyed for chemistry which is applicable on a large scale, e.g. cheap reagents, useful catalysts, reactions which could be carried out in water or under phase transfer conditions and reactions which use common bases such as sodium carbonate instead of organolithium reagents.

Indeed, to follow a direct path from laboratory data to the pilot plant requires either information that is often unavailable or scientific and engineering judgements beyond those normally considered possible or desirable.

IV.   The Role of Chemical Development

The development chemist's job is the design of a process with the following qualities:
- a)   The process must be safe.
- b)   It should be cost efficient. Scale-up processes do not involve only technical decisions and compromises. The selected compromise always has an economic aspect since it is never possible to establish exactly what an industrial process should be. There are always restrictions of time and money.
- c)   The process must be ecologically sound. The costs involved in meeting environmental, occupational health and safety, and consumer product safety requirements are so significant that they can, and often do, become prime considerations in whether to proceed with process development and scale-up.
- d)   The process must be reproducible and robust.
- e)   The chemistry must fit the existing equipment. To design a process with the desired qualities, the development chemist must understand the chemistry and the mechanisms. Process development and scale-up require multidisciplinary cooperation between organic chemists, analytical chemists, process engineers, pharmacologists, toxicologists, clinicians and marketing specialists. Chemical engineers are involved in plant trials, plant modification or new plant construction. Chemical engineers should be an essential part of the development team, as they bring a different point of view. A chemical engineer involved in bulk chemical production, envisages the pilot plant as a purpose-built plant designed to test a process, contrary to the development chemist, who sees the plant as a logical extension of a typical laboratory set-up of three-neck flask with addition funnel, stirrer and condenser. The cooperation of chemists and chemical engineers is vital to the success of scale-up, particularly the later stages of introduction of a process to full plant scale. Chemical engineers teach chemists the importance of heat transfer, mass transfer and kinetics, and other subjects with which the chemist may be less familiar.

When a development chemist begins his task of synthesizing a target compound, the following information is usually available: a) A research method for making gram quantities. b) Possibly a sample. c) Possibly some analytical methods and cleaning methods. d) A patented synthetic route which may not be, and usually is not the best way to prepare the compound. It is simply the way a medicinal chemist decided to prepare several analogs of a class of compounds for testing. Therefore it is vital to take a long term approach and seek the best route at the start. The rule of thumb is that the shortest route is usually the best. A short route means shorter plant time which translates in lowering capital cost. To choose a synthetic route it is important to generate several paper routes. Talk to colleagues and consultants, search the literature for similar compounds. Finally decide on a few which appear to have a high chance of success, but also try some speculative routes which could be beneficial even if there is little literature precedent.

In order to design new synthetic routes to produce inexpansive products, it is important to choose the correct raw material. The chemist working on the development and scale-up of a new chemical entity, needs to have a wide knowledge of the fine chemicals market so that he knows not only what is available but also how much it costs. Often the only reason a chemical is not available in bulk is lack of demand.

V.   Optimization in Chemical Development

Chemical development involves not only synthetic route selection, but also optimization, scale-up and further improvement of the synthetic method until an efficient process is obtained. The development chemist's task is to convert a laboratory synthetic route, which used expensive reagents and very dilute reactions to a robust manufacturing process. In doing so the development chemist should scale-up the process to 1-30 Kg. Scaling up a process allows the development chemist to test the process at the pilot plant level and to identify problems. He should learn how to handle reactants, intermediates and waste streams. How are isolations done? Are filtrations fast or slow? What kind of filtering equipment is needed?

To scale-up a process from the laboratory to the pilot plant the development chemist must have an appreciation of chemical engineering principles. Understanding of heat transfer and mass transfer is essential in the successful scale-up of a new process. To control the temperature with respect to time is one of the most important aspects of chemical engineering. Reaction temperatures must be controlled in order to ensure selectivity of the process, reproduce results accurately and to prevent thermal runaways. Heat transfer is not only important in the reaction but in the work up as well. For example, to control the temperature during exothermic neutralizations, crystallizations, and distillations. In many chemical processes the rate of external heating may not be important, but the rate of external cooling can be very critical when exotherms take place.

Reactions with low volume are most difficult to control due to the fact that the removal of heat is proportional to contact surface area as well as ΔT. Heat transfer can be effected by a number of factors such as rate of agitation, turbulent or laminar flow, viscosity of reaction medium, density, temperature, shape and surface of vessel, exothermicity of reaction and phase changes. The heat transfer can be controlled by using the correct vessel for the batch size. The material of construction (stainless steel, glass etc.) as well as the design and agitator are important. Heat evolved is proportional to number of moles of reactant(s), and is therefore proportional to volume of solution $(r^3)^5$. Removal of heat is also proportional to surface area $(r^2)^5$. Therefore as vessel size increases, volume-to-surface ratio also increases. This means that heat transfer becomes more difficult. The development chemist should perform a risk assessment on the process he is scaling up. For example, can a process temperature be controled by the existing cooling system? What temperature could be attained after the runaway of the desired reaction? At which moment does cooling failure have the worst case consequences? How fast is the runaway of the desired reaction?

Mass transfer is also important in successfully scaling up a process. Mass transfer (which is not the same as agitation) becomes very important in a reaction with more than one phase and in work-up and purification. For substances to react they must come into contact. In a two-phase system, mass transfer is affected by the rate of diffusion to and across the interfacial boundary and the rate of diffusion of products from the reaction zone. Mass transfer is affected by the size of the interfacial area or surface as well.

A large vessel is not identical to a small vessel in most of the mixing and fluid mechanics parameters that may be involved in determining process results. To scale up from a small laboratory vessel to a large vessel in the pilot plant or in production, understanding of some of the principles of fluid mixing is required. This ensures consistency in purity and yield of the product. The development chemist knows all too well that agitation can be fatally different in the plant. For example, laboratory glassware is spherical where plant equipment is more cylindrical, and plant vessels are baffled resulting in better mixing. Good mixing is vital in obtaining good mass transfer, especially when we are dealing with a viscous medium. Many times, engineers spend a lot of time designing an agitator for a particular process. Some type of agitators are: The impeller, which consists of three radial and curved blades assembled axially with 1-2 baffles[6]. The twin agitator, which is made up of several two-bladed wheels rotated ninety degrees and assembled with or without baffles[6]. The anchor, which is assembled axially without baffle or with one thermopocket[6]. The loop agitator, which is a tubular gate agitator6. The uniflow axial turbine, which is made up of pitched blades to give high axial flow and low shear[6]. The pitched turbine is a six-bladed propeller assembled axially with 1-4 beaver-tail baffles or eccentrically without baffle6. The radial turbine is made up of flat bladed to give high radial flow and high shear[6]. Finally the disc turbine is a turbine wheel with six radial blades assembled axially with or without 1-4 beaver-tail baffles[6].

To mimic the laboratory mixing conditions, a vessel must be designed to have the same Reynolds number (Nre). The Reynolds number can be obtained from the following formula[7]:

$$Nre=Dv\rho/\mu$$

(1)

This formula takes into account the diameter of the vessel (D), the velocity of the fluid (v), the density (ρ) and the viscosity (μ). The Reynolds number has no dimensions. For high viscosity applications one can see that we need large scale diameter agitator and low speed. On the other hand for low viscosity fluids the diameter of the agitator may be as low as one third the vessel diameter and the speed high. In scale-up the ratio of impeller diameter to vessel diameter is an important factor.

To select a particular type of reactor in any stage of scale-up, it is important to identify your purpose and/or goal. For example, for the small scale manufacturer, few criteria are as consequential as flexibility with respect to operating conditions and product demand. The product market might not justify a massive research program or it might be ideal to produce several compounds out of the reactor. The economics of the overall process are often little influenced by the cost of the reactor. The ability of a chosen reactor to optimize product selectivity, or maximize productivity, is far more important in practice than the minimization of the reactor cost.

Control of temperature, and mass transfer is of great importance in the design or choice of reactor in scale-up. The rates and equilibria of many reactions are profoundly affected by temperature. Accordingly, so are side reactions, by-product formation, yield, selectivity, and so on. The great variety in reactor types and the fact that no single type is suitable for all purposes makes it necessary to establish criteria for selection. One such criterion is the type of mixing needed for the process in question. The role of mixing in all these various processes may not be completely understood. This makes a qualitative understanding of the differences between the pilot plant and the laboratory and their possible effects much more important to understand.

The purpose of scale-up is to actually observe, measure, analyze, and record data.

VI.   Process Safety

No process can be made one hundrend percent safe. However, one can identify hazards and, if the hazard is severe, avoid such a chemical or process, or if the hazard is moderate, take precautions that minimize the risk. Safety hazards can be divided into four categories: thermal instability, toxicity, flammability, and explosiveness. Typically the in-house safety laboratory measures the decomposition temperature of all reagents, intermediates, solvents, distillation residues and evaporation residues, and any exotherms associated with the decomposition, so that one stays well below these temperatures in the process. The heat of reactions is measured to ensure adequate cooling capacity of the reactor before scaling up a reaction. This minimizes the risk of runaway reactions.

The LD_{50} and carcinogenicity of compounds are measured as well. For known compounds, one can look up the data in manuals, such as Sax's Dangerous Properties of Industial Materials. For new compounds, the tests are done in-house, both the LD_{50} and the Ames test. There is no hard rule for what is too toxic in the plant, but typically the rule of thumb may be following table I. Solvents that are carcinogens are avoided, e.g., chloroform, carbon tetrachloride, dioxane, benzene and hexamethylphosphoric triamide[8].

Table I. Toxicity of Chemicals

LD_{50} (mg/Kg)	Toxicity
<0.5	Dangerously toxic
0.5-50	Seriously toxic
50-500	Very toxic
500-2000	Moderately toxic
2000-15000	Slightly toxic
>15000	Low toxicity

Another safety hazard is flammability. It is a concern mainly with solvents that are volatile. One can easily find the following information: ignition temperature, flash point, vapor pressure, vapor density and the mixture of solvent and air that is flammable. The lower and the wider the latter value, the more hazardous the solvent, since more likely one will encounter the conditions for flammability in the plant.

Again, there is no hard line between what is safe and what is not, but solvents with flash point of less than $-18°C$ are not used. A solvent with a flash point above $-18°C$ can be used but special precautions must be taken such as grounding of all drums and reactors to prevent buildup of electrostatic electricity and excluding all other sources of ignition. All other solvents (toxic or flash point $<-18°C$) must be replaced. For example, chlorinated solvents (chloroform, carbon tetrachloride), which are carcinogenic, can be replaced with ethyl acetate. Solvents such as pentane (flammable) or hexane (forms electrostatic charges) are replaced with heptanes. Benzene (a carcinogen) is replaced with toluene or xylene. Diethyl ether is replaced with t-butyl methyl ether and dioxane with tetrahydrofuran.

The development chemist must also determine if these changes adversely affect the reaction chemistry. Often a change in solvents can dramatically alter the outcome of a reaction.

Another safety issue is explosiveness of compounds. This factor is evaluated by the hazards laboratory with tests such as the dust explosion test and the hammer test. Many times we can predict the explosiveness of compounds by looking at their structures. Compounds containing weak bonds between heteroatoms such as in peroxides, hydrazines, halamines and hydroxylamines as well as compounds that can eliminate small, stable molecules like N_2, O_2, NO, or NO_2 (e.g., diazonium, ozonides, nitroso, and nitro compounds) are potentially explosive.

VII. Environmental Safety

In the last two decades environmental considerations and challenges have become a primary concern and significantly impacted the business and technical community. As long as processes did not create any local nuisances, such as bad odors, environmental considerations were not part of the design of processes or manufacturing philosophy. The designers of a process considered the safety of personnel from the point of view of catastrophies, but did design into the systems safeguards from chronic as well as acute exposures to environmental insults. The cost involved in meeting environmental, occupational health and safety, and product safety are so significant that they often become prime considerations in whether to proceed with process development and scale-up.

Beginning with Earth Day 1970 the casual approach to pollution has changed drastically. The process chemists and engineers must consider environmental aspects of any process. Public pressure, as well as regulations, require that we minimize waste discharges and design for zero environmental insult. To the development scientist this means that an ideal process must be environmentally acceptable. Nowdays ecology is a number two priority, right after safety. During the design of a process, the development scientists must address all three environmental sinks: land, air and water. To meet this goal the development scientist must ask several questions: 1) can we recycle solvents or reagents? 2) can we minimize the use of solvents or reagents, since waste streams are affected? 3) can we avoid toxic solvents and reagents. 4) can we avoid low boiling solvents since they can easily escape into the atmosphere. 5) can we use "clean" chemistry to avoid the generation of hazardous waste?

Solvents in Class I (table II) should not be used in manufacturing drug substances, excipients, or drug products because of their unacceptable toxicity or their deleterious environmental effect. However, if their use is unavoidable in order to produce a drug product with a significant therapeutic advance, then their levels should be restricted as shown in table II, unless otherwise justified[9].

Table II. Class I solvents

Solvent	Concentration limit	Concern
Benzene	2ppm	Carcinogen
Carbon tetrachloride	4ppm	Toxic & Environmental Hazard
1,2-Dichloroethane	5ppm	Toxic
1,1-Dichloroethene	8ppm	Toxic
1,1,1-Trichloroethane	1500ppm	Environmental Hazard

Solvents in table III should be limited in pharmaceutical products because of their inherent toxicity. Potential adverse effects (PAE) are given to the nearest 0.1 mg/day, and concentrations are given to the nearest 10 ppm. Solvents with low toxic potential, class III, table IV, may be regarded as less toxic and of lower risk to human health.[10,11].

Table III. Class II solvents

PDE (mg/day)	ppm
Acetonitrile	410
Chlorobenzene	360
Chloroform	60
Cyclohexane	3880
1,2-Dichloroethane	1870
Dichloromethane	600
1,2-Dimethoxyethane	100
N,N-Dimethylacetamide	1090
N,N-Dimethylfomamide	880
1,4-Dioxane	380
2-Ethoxyethanol	160
Ethyleneglycol	620
Formamide	220
Hexane	290
Methanol	3000
2-Methoxyethanol	50
Methylbutyl ketone	50
Methylcyclohexane	1180
N-Methylpyrrolidone	4840
Nitromethane	50
Pyridine	200
Sulfonane	160
Tetralin	100
Toluene	890
1,1,2-Trichloroethene	80
Xylene	2170

Table IV. Class III solvents

Solvent	Solvent
Acetic acid	Heptane
Acetone	Isobutyl acetate
Anisole	Isopropyl acetate
1-Butanol	Methyl acetate
2-Butanol	3-Methyl-1-butanol
Butyl acetate	Methylethyl ketone
Tert-Butylmethyl ether	Methylisobutyl ketone
Cumene	2-Methyl-1-propanol
Dimethyl sulfoxide	Pentane
Ethanol	1-Pentanol
Ethyl acetate	1-Propanol
Ethyl for-mate	Propyl acetate
Formic acid	Tetrahydrofuran

VIII. Conclusion

Chemical development provides key deliverables to other departments; that means that the development chemist works under the restrictions of deadlines (the drug substance is always needed yesterday), trying to find ideal process qualities (see summary), and performing targeted research that is unpredictable. The development scientists work with the best processes available at the time, not with an ideal process.

The prediction of process performance in pilot plants and production plants has been part of the science of chemical engineering for many years. The goal is to model and predict what willhappen while conducting the minimum number of experiments. Scale-up studies should involve modeling relevant phenomena, not the study of miniaturized commercial systems. The miniplant concept has been developed to predict the performance of large-scale equipment from data generated by the low-cost experiments in small equipment, rather than relying solely on fundamental data.

A number of rules of thumb we have found are: Liquid and vapor handling are easier than solid handling. Recycle streams usually build up impurities. An adiabatic reactor is usually easier to scale-up than a jacketed cooled reactor. Heat transfer varies with construction material of reactor. Separations are lengthy and often surprising results are seen. Extractions are lengthy and must be piloted. Distillations often do not need piloting. Solid handling, crystallizations and drying are processes that must be piloted before full production.

Once King Ptolemy of Egypt asked Euclid whether he could make his geometry simpler, so that Ptolemy himself could learn it with less effort. Euclid repried; "There is no Royal Road to Geometry". This is certainly true in chemical development, there is **No *Royal Road to Scale-up*.**

References

1. Scrip Magazine, 2669, August 2000.
2. PharmaBusiness, P.62, March/April 2001.
3. Chirasource USA, Symposia in Philladelphia, January 2000.
4. Pharmaceutical Executive, P.164, May 2000.
5. K. Denbigh, Chemical Reactor Theory, P.2-23, Cambridge at the University Press, 1965.
6. O.Levenspiel, Chemical Reaction Engineering, P.309-311, John Wiley and Sons, Inc., New York, 1962.
7. J.G. Bralla, Design for Manufacturability, P.500-520, TWI press, 2001.
8. EPA, Chlorinated solvents, 1999.
9. J.A. Riddick, W.B. Burger, Organic Solvent, John Wiley and Sons, P.255-258, 1986.
10. EPA, Exposure limits, 1999;
11. A. Gordon, R.A. Ford, The Chemist's Companion, John Wiley and Sons, 1973.

Chapter 2

Design and Development of Practical Syntheses of MRSA Carbapenems

Edward J. J. Grabowski

Merck Research Laboratories, Department of Process Research, Merck & Company, Inc., P.O. Box 2000, RY 800-A231, Rahway, NJ 07065

Process research efforts relative to the design of a practical synthesis of a structurally complex MRSA carbapenem (**1**) culminated in the establishment of two distinct syntheses. The first is based on a unique Stille coupling between a carbapenem enol triflate and a stannatrane appended via a methylene spacer to a complex and extended heteroaromatic side chain. The second is founded upon a novel π-allyl palladium reaction to incorporate the entire complex side chain. Details on the development of both methods are described.

The design of practical syntheses of carbapenems remains amongst the most formidable of challenges within the realm of process research, despite the remarkable achievements that have been made in the field in the last generation.

The commercial availability of a number of key carbapenem building blocks along with the spectacular growth in synthetic methodology has enabled medicinal chemists to explore structure-activity relationships that would have been considered beyond reality when the field first began. Since the days of imipenem, Merck has remained very active in the field, most recently seeking compounds effective against hospital-acquired MRSA infections. Based on these exceptional efforts, a number of remarkable compounds have come into development.[1a,b,c,d] The most recent candidate which required our attention is compound 1, which has a number of unusual attributes. In addition to having the β-methyl group in the carbapenem nucleus, the compound has an elaborate side chain, which is larger than the carbapenem portion of the molecule! This side chain consists of a naphthosultam portion linked through a DABCO nucleus to an acetamide. In-house it is considered as belonging to the cationic zwitterion class of antibiotics.

1

Probably the most unique feature of 1 is the presence of the methylene spacer between the 2-position of the carbapenem nucleus and the nitrogen of the naphthosultam. Because of concerns that the residues of these carbapenems might illicit an immune response once they acylated a blood borne protein, the methylene spacer was put between the two components of the molecule to allow it to fragment after acylation. Thus, the side chain would become available for normal metabolic-elimination processes, and the bound residue of the carbapenem would be too small to effect the immune reaction. Internally, this became knows as the 'releasable hapten hypothesis', and served to generate a number of exciting new compounds. (Figure 1).[1a]

The starting point for the design of any practical synthesis in a process group is the medicinal chemistry route. Most often the latter is designed for versatility not efficiency. However, it is always a good source for excellent ideas for designing a practical route, and it is an excellent source for information on the characteristics of key intermediates. The medicinal route to 1 (Figure 2) built up the key 2-hydroxymethyl carbapenem from the commercially available protected

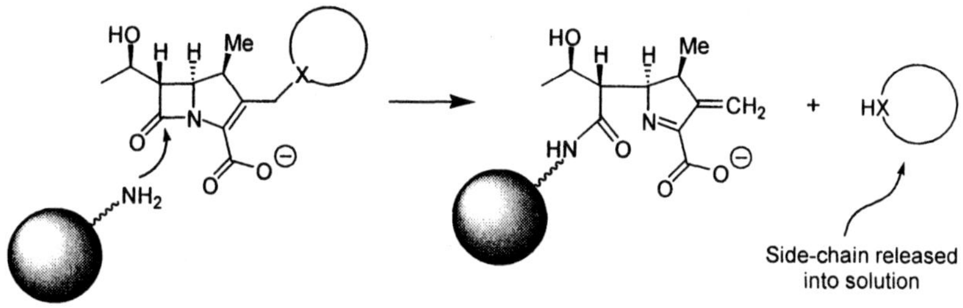

- Avoid immune recognition by coupling β-lactam ring opening with a chemical fragmentation to release the immunogenic hapten
- Chemical fragmentation well known among cephems

Figure 1. The "Releasable Hapten" Hypothesis

acetoxyazetidinone in twelve steps in 20-27% yield based on available literature chemistry.[2] With the allyl protecting groups, the 2-hydroxymethyl carbapenem intermediate was not a well-behaved compound with regard to stability and crystallinity, and clearly one we would want to avoid. The subsequent steps (Mitsunobu insertion of the sultam, DABCO-acetamide addition and deprotection) proceeded in 32% yield for an overall yield of 6-12% for the synthesis, and produced decigrams of product. While we could live with the latter part of the synthesis for the production of material for safety assessment and Phase I clinical trials, we needed a better route to the 2-hydroxymethyl carbapenem, hopefully in a form in which we could easily handle it.

Dr. Nobu Yasuda, Mr. Chunhua Yang and Mr. Ken Wells came up with the idea of combining some chemistry developed for an earlier carbapenem with the tri-n-butyltinmethanol reagent pioneered by Kosugi.[3] The enol-triflate shown in Figure 3 was prepared from the commercially available TES-diazoazetidinone via the rhodium-mediated insertion reaction followed by treatment with triflic anhydride and base.[1] This intermediate was previously prepared *in-situ* and used successfully used for a variety of Suzuki coupling reactions. In this application it was treated with the tinmethanol reagent under typical Stille conditions to give the desired hydroxymethyl carbapenem in 45-55% yield. All was not perfect with this reaction, as side products arose from reduction and butylation, and HMPA was a prerequisite for even this modest yield. Because of the change of protecting groups from allyl to p-nitrobenzyl (our favorite for carbapenem synthesis), the hydroxymethyl compound was crystalline, and could be stored for future needs.[4] With additional development we were able to prepare decakilograms of the hydroxymethyl intermediate for our program needs.

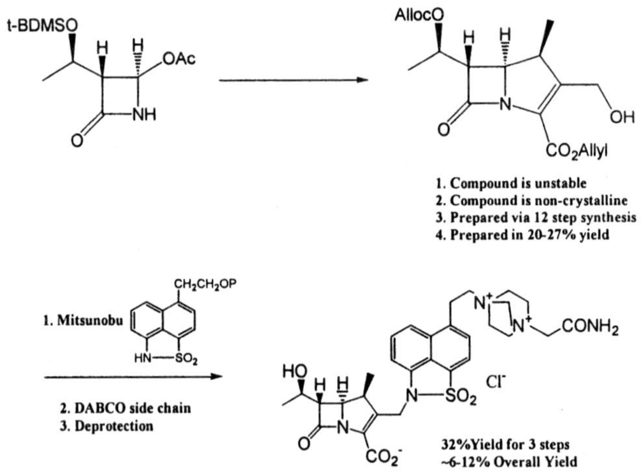

Figure 2. Medicinal chemistry route to the MRSA carbapenem.

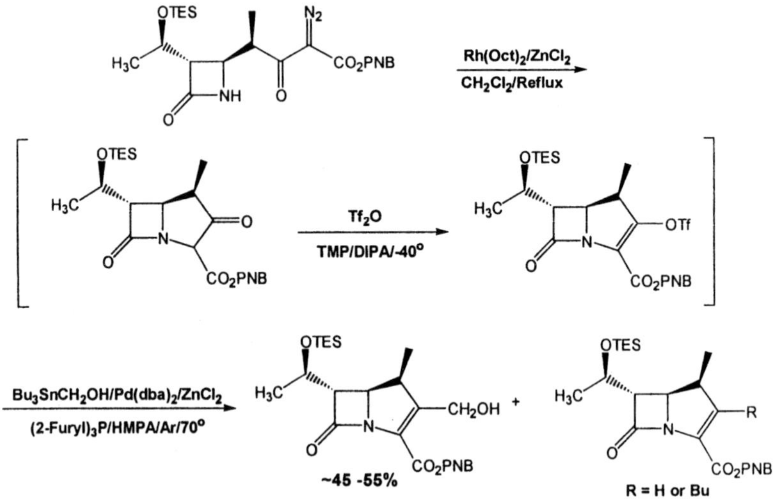

Figure 3. Application of Bu₃SnCH₂OH chemistry (Kosugi 1985).

With the key intermediate in-hand, we were able to complete the synthesis and provide enough material for safety assessment, formulation studies and Phase I clinical studies employing the π-allyl palladium chemistry noted below.

While this solved our immediate problems, the long-range problem of a practical synthesis of this target remained, and it was time to regroup and address this issue. As noted in the introduction, designing practical syntheses of these compounds is among the most formidable of problems that a process research chemist faces. In fact, there are really three problems hidden within the context of this one program: a practical synthesis of the side chain (which has been ignored until this point), elaboration of an appropriate carbapenem system and coupling to the side chain to afford penultimate product, and, finally, deprotection. While the last just appears as 'one arrow' in a reaction scheme, it is in fact a formidable step as one needs to isolate a relatively unstable, water soluble product from a complex reaction mixture in a stable form that is pure, sterile and pyrogen free. With these objectives in mind, we expanded the program team to be able to address these issues simultaneously.

Drs. Guy Humphrey and Ross Miller and Mr. David Lieberman sought to design and demonstrate a practical synthesis of the naphthosultam side chain. The medicinal chemistry route to this intermediate was based on 1-(2-hydroxyethyl)naphthalene, a compound only available from specialty supply houses in limited quantities. Clearly this would not suffice, and an alternate starting material was needed. Dr. Miller recognized 1-methylnaphthalene as the most likely candidate for this need. It is available in bulk for a few dollars/kg, and should be amenable to a similar sequence of electrophilic reactions as used by the medicinal chemists to build up the sultam ring.

The first approach to the side chain synthesis is shown in Figure 4. Chlorosulfonation of the naphthalene proceeded in good yield to give the 4-chlorosulfonyl compound with excellent regioselectivity. Nitration gave a 3:1 mixture of desired vs undesired regioisomers, which was purified to the desired 1,4,5-isomer in 59% yield with an MTBE wash. Benzylsufonamide formation and an internal SnAr2 displacement (a reaction from the medicinal group) afforded the desired 1-methylnaphthosultam in workable overall yield and excellent purity.

Fortunately, we always ask our hazard evaluation group to look at any chemistry destined for the preparative scale. To our dismay, the nitrosulfonyl chloride intermediate proved frighteningly shock sensitive and thermally unstable! This necessitated an immediate revision in the chemistry to avoid this intermediate (Figure 5). The initial chlorosulfonyl intermediate was converted to its N,Ndiethylsulfonamide and then nitrated. The same mixed regiochemistry prevailed, but workup afforded the desired nitrosulfonamide in comparable yield to the initial route and in excellent purity. Transfer hydrogenation and solvolytic closure provided a route to the methylsultam suitable for the preparative scale and free of dangerous intermediates.

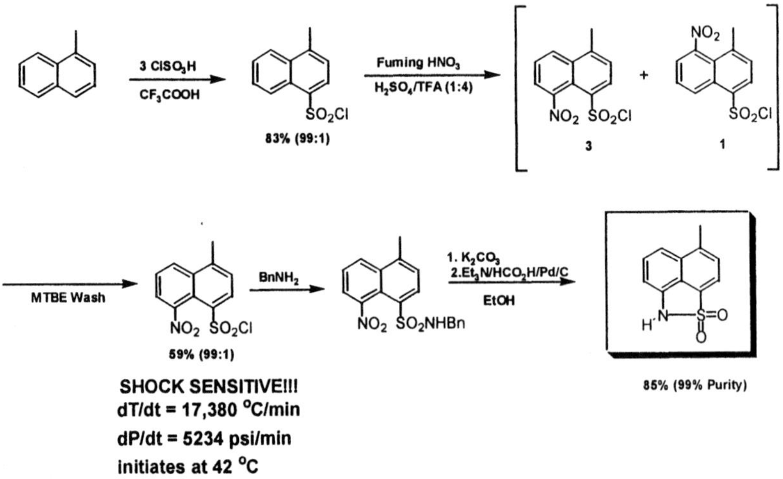

Figure 4. First side chain synthesis from 1-methylnaphthalene.

We now faced the problem of adding the hydroxymethyl group to the system. The dianion of the 1-methylsultam underwent ready reaction with a variety of electrophiles, and in generally good yields, save for those electrophiles that afforded direct hydroxymethylation. In terms of overall yield and operational simplicity, carboxylation and borane reduction proved to be the best method for preparing the hydroxyethylsultam. Subsequent triflate activation of the hydroxyl group and displacement with the DABCO acetamide triflate salt afforded the complete side chain in a synthesis readily amenable to scaleup. Attesting to the viability of this route, over thirty kilograms of the side chain were prepared in our preparative laboratory. The overall route if summarized in Figure 5.[5]

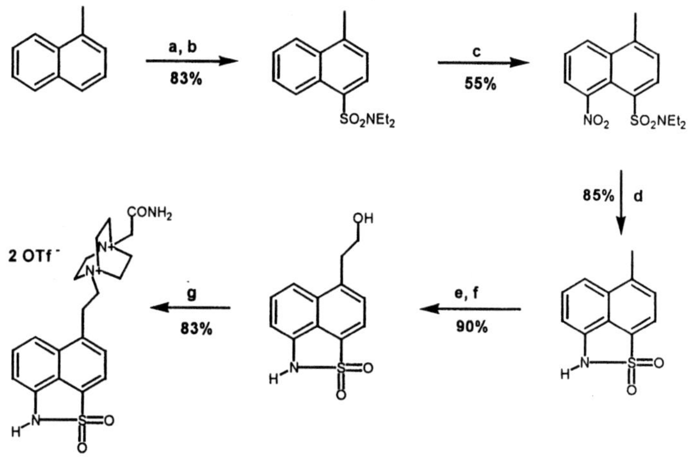

a) 2 eq. ClSO$_3$H; b) HNEt$_2$, IPA; c) HNO$_3$,
H$_2$SO$_4$; d) KO$_2$CH, 5% Pd/C, HCl; e) LDA,CO$_2$; f)
NaBH$_4$, BF$_3$OEt; g) Tf$_2$O, DAT

Figure 5. Summary of the side chain synthesis.

While Guy Humphrey's team was dealing with the issues of the side chain, Drs. Nobu Yasuda and Mark Jensen were studying new methods for introduction of the side chain. Remembering that the enol triflate in Figure 3 underwent a Stille coupling with the tin methanol reagent, numerous Stille couplings were attempted with Bu$_3$SnCH$_2$-Sultam, wherein the sultam was attached via its acidic nitrogen and the sultam was varied from the simple desmethyl model, to its hydroxyethyl analog to finally the full side chain. In none of these attempts was significant coupling observed.

Concerned that the tin-methylene bond might be too strong for an effective Stille coupling in these examples, they proceeded to recast the tin reagent as its stannatrane analog based on the work of Vedejs, wherein it was thought that the nitrogen backbonding to tin in the methyl analog rendered the methyl more readily transferable under Stille conditions.[6] Two stannatrane-based reagents were prepared, and both reacted with the enol triflate to afford excellent yields of the desired coupled products. Employing the stannatrane with the entire side chain in place, a 98% yield of the TES-penultimate product as its ditriflate salt resulted from this coupling (Figure 6).[7] The stannatrane was completely recovered as its chloro analog, ready for recycle.

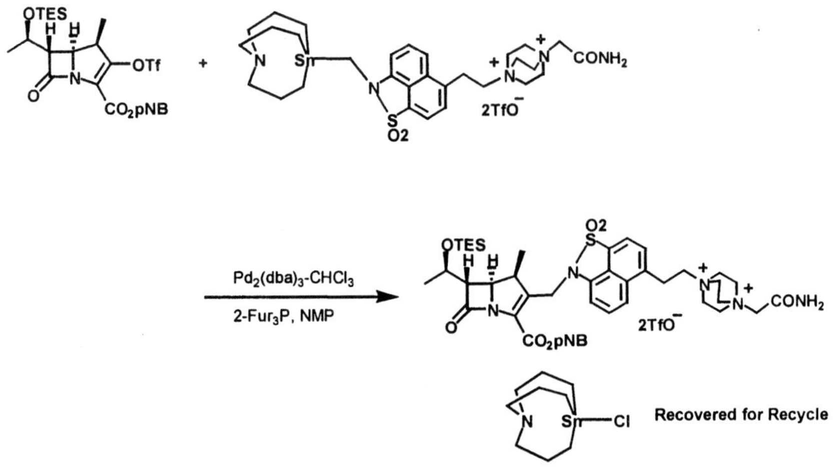

Figure 6. Stannatrane coupling reaction.

Subsequent deprotection (removal of TES under acidic conditions and the PNB-group under reductive conditions with controlled pH) afforded the desired carbapenem as before.[7]

While the stannatrane chemistry was under study, Guy Humphrey had a wonderful idea. Would the hydroxymethyl compound used for our original large scale synthesis of the carbapenem target (See Figure 3) be amenable to π-allyl palladium chemistry once a suitable leaving group was attached to the hydroxymethyl group? To our knowledge, this type of chemistry had not been applied in carbapenem synthesis. With the protected hydroxyethyl version of the sultam side chain as its potassium salt and employing palladium acetate-triphenyl phosphine (1:3 with 3 mole% Pd) to mediate the reaction, poor conversion was noted with acetate as the leaving group. However, with carbonates or phosphate as the leaving group yields in the 85-90% range resulted in THF or acetonitrile at reflux. The same chemistry obtained even when the hydroxyethyl group of the sultam was left unprotected. In the absence of the palladium catalyst no coupling took place.

Encouraged by these preliminary results, Guy and his team proceeded to develop this reaction. The results are illustrated in Figure 7. Despite the use of relatively

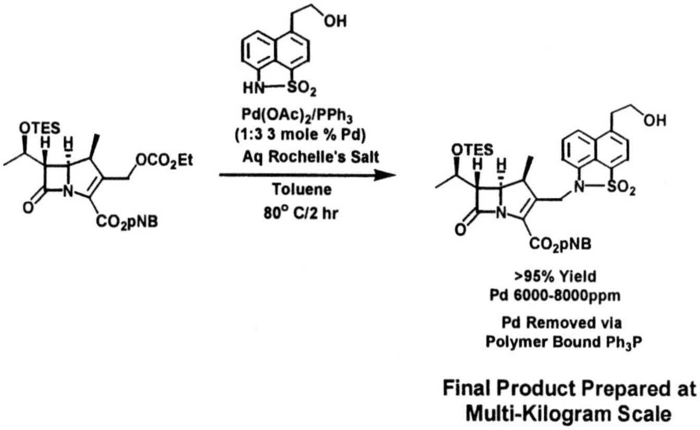

Figure 7. Developed π-allyl palladium chemistry.

harsh conditions (toluene-aq. Rochelle's salt at 80°), the reaction proceeded in over 95% yield with little destruction of the carbapenem ring, and it was successful at the multikilogram scale. However, there was some bad news. Most of the palladium used in the reaction ended up in the product, and treatment with a phosphine containing resin was necessary to remove it.

This exciting result prompted two key questions. Could the entire sultam side chain be incorporated via the π-allyl palladium chemistry? Certainly there would be major changes in overall reaction conditions employing the dicationic side chain. The second question was even more intriguing. Could the desired carbonate be made by a simple and scaleable synthesis? Although we had run the Stille tinmethanol coupling on a kilo scale, we wanted no more of it. It had served its purpose in allowing the safety and clinical studies to start, and in giving us time to develop more attractive alternates.

Joining the effort at this time were Drs. Kai Rossen and Phil Pye, specifically to work on a practical carbonate synthesis. In thinking about how to approach this problem (Figure 8), their thoughts led back to the acetoxyazetidinone, which, in fact, was defined as a viable carbapenem synthon by Dr. Paul Reider and yours truly back in the early 1980's.[8,9] In the intervening years it has become an article of commerce. The obvious key problem in this effort would be to establish the β-methyl stereochemistry. While solutions exist, none are applicable because of the extra methylene spacer present in **1**. Kai and Phil were pleased to note that 1-hydroxy-2-butanone was available in ton quantities from suppliers in Europe. They embarked on an extensive study of the reactions of protected enolates of this ketone with the acetoxyazetidinone, which is beyond the scope of this discussion. The conclusion from their work was that reaction of the titanium enolate derived from the iso-butylcarbonate of the ketone with the acetoxyazetidinone afforded the desired β-methyl adduct with 95:5 selectivity (Figure 9). The simplicity and selectivity of this reaction is unprecedented in my experience.[10]

Since the azetidinone starting material was protected with the tBDMS-group, and we could not remove that without destroying the fully elaborated carbapenem, its removal under acidic conditions was required at this point. Our initial concerns turned to joy when we realized that the resulting hydroxyethyl azetidinone was not only wonderfully crystalline, but it upgraded to ~99.5% in favor of the desired diastereoisomer in 63% overall yield! Reprotection with TES chloride, reaction with PNBoxalyl chloride, phosphite cyclization and recrystallization afforded the desired carbonate in 52% overall yield. In a single preparative lab demonstration of this chemistry, seven kilograms of intermediate was made to support our needs (Figure 9).[10]

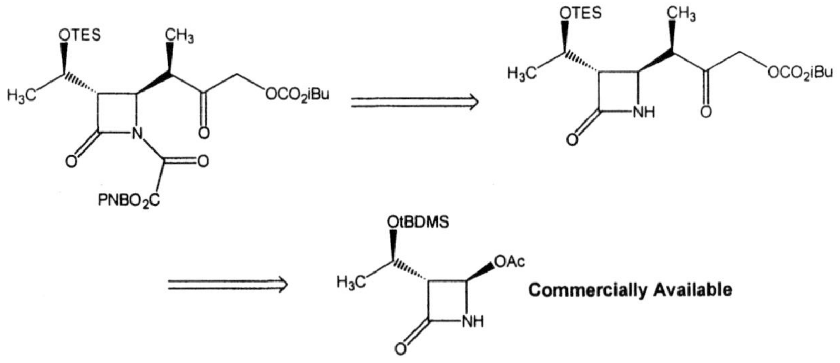

Figure 8. Carbonate retrosynthesis

With an ample supply of the carbonate in hand, focus turned to the ultimate development of the π-allyl palladium chemistry. Since Guy Humphrey's team intended to couple the entire side chain in this approach, it was important to minimize severe reaction conditions during the coupling. While triphenyl phosphine required a reaction temperature of 80° for completion of the reaction in two hours, such could be done most efficiently employing trifuryl phosphine in its place. Even more exciting was the fact that coupling could be done at 25° in one hour employing triethyl phosphite in place of the phosphines. This set the stage for the ultimate coupling of the whole side chain.

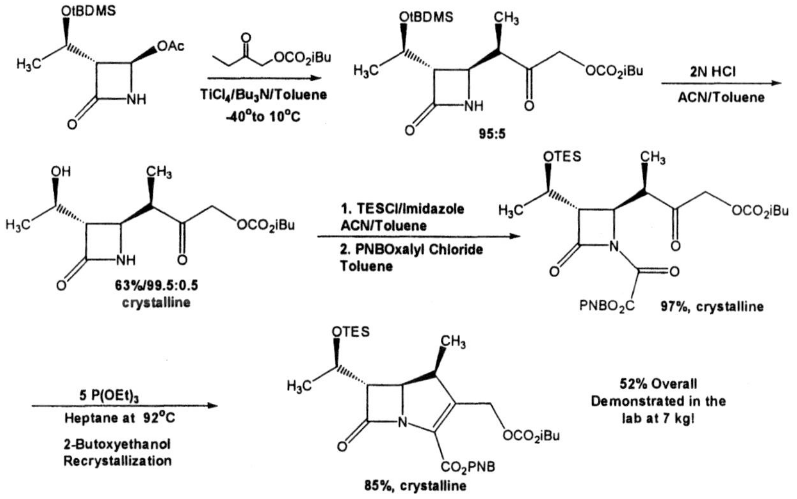

Figure 9. Carbonate synthesis

Numerous changes were necessary to effectively couple the full side chain to the carbonate as defined by Dr. Ross Miller. A switch to NMP, with lutidine as the base, afforded complete conversion of the carbonate to the desired coupled product (Figure 10). Isolation of the TES-penultimate product was accomplished by the simple addition of water-isopropanol to the reaction mixture. The resulting product was isolated in 97% yield and 99% LC purity with 95 ppm of residual palladium on a three kilogram scale.[10]

This set the stage for the final deprotection, and Ross took charge. For years we had always isolated cationic carbapenems as chlorides, sulfates or phosphates, and in the present case none of these salts was truly satisfactory for all of our needs. Ross began to examine a series of new anions for salt formation, and discovered that aryl sulfonates in general formed excellent salts. Of these the benzenesulfonate salt proved best, affording an exquisitely crystalline and stable salt, that enabled isolation of final product without resort to the traditional isolation techniques (resin columns, pod extractors, wiped film

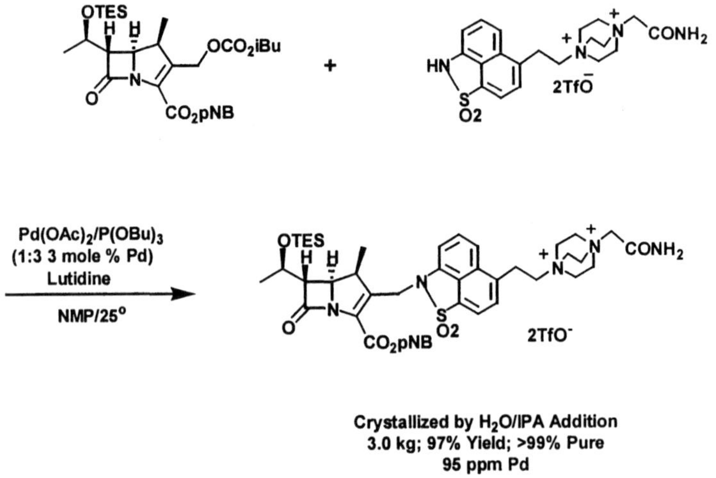

Crystallized by H$_2$O/IPA Addition
3.0 kg; 97% Yield; >99% Pure
95 ppm Pd

Figure 10. π-Allyl palladium convergent process.

evaporators, etc.) which we normally employ with this class of compounds. The resulting deprotection process is so simple to run that hundreds of grams of product can be made in a simple laboratory run! (See Figure 11)[10] Although the benzenesulfonate salt is too insoluble for iv administration, it will dissolve in dilute saline or it can be converted to the chloride by exchange in the presence of a quaternary ammonium ion resin on the chloride cycle in water.

The story might end here, save for the thoughts of Dr. John Chung, another of the chemists in our Process Group. In the best traditions of a process group, John thought to build further on what was already established, while at the same time simplifying the overall process. Going way back to the original tin methanol Stille coupling with the enol triflate, John reasoned that if the triflate would react with the nitromethane anion, the resulting nitromethyl compound could serve as a substrate for the π-allyl palladium chemistry. Thus, with one step we could tie the established enol triflate chemistry, which had been used on the pilot scale for earlier carbapenem analogs, to the π-allyl palladium chemistry and thereby

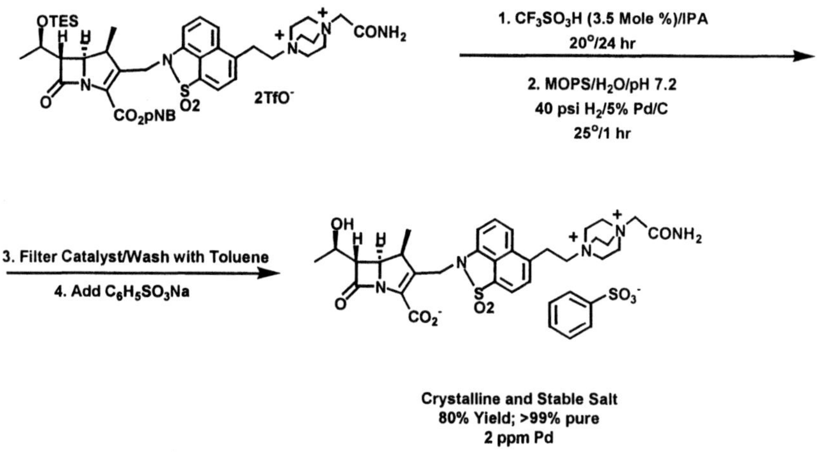

Figure 11. Conversion to final product.

effect a new synthesis of the target that could compete with the existing syntheses. In principal the chemistry worked; in practice it suffered from severe problems. As noted in Figure 12, reactions of the enol phosphate (a commercially available compound!) suffered from principal attack of the nitromethane anion at the heteroatom instead of the C-2 carbon. This situation was somewhat remedied with the enol triflate where the selectivity was at least 5:1 in favor of attack at carbon instead of sulfur. However, this only occurred in a solvent system consisting of 1:1 HMPA or DMPU/nitromethane with tetramethyl-guanidine as the base. The resulting nitromethyl adduct did undergo the desired π-allyl palladium chemistry, but the overall yield was only ~35%. Clearly this reaction would not compete with the existing chemistry. Turning a sow's ear into a silk purse, John then explored the generality of the method: reaction of the enol triflate of a β-keto ester with nitromethane anion followed by π-allyl palladium chemistry with soft carbon, nitrogen and sulfur nucleophiles. It proved to be quite general, and a report of his work has been published.[11]

Of the three methods that we have developed for this target, the stannatrane chemistry clearly has that certain *je ne sais quoi*! However, ultimately it would have to deal with the synthesis and recovery of the stannatrane metallocycle, the concern with handling tin compounds in a production setting and concern with tin in the final product. The π-allyl palladium route would be the route of choice should we pursue this compound to the manufacturing stage. It already has all of the key elements of a good manufacturing process, and there are areas for additional development and simplification.

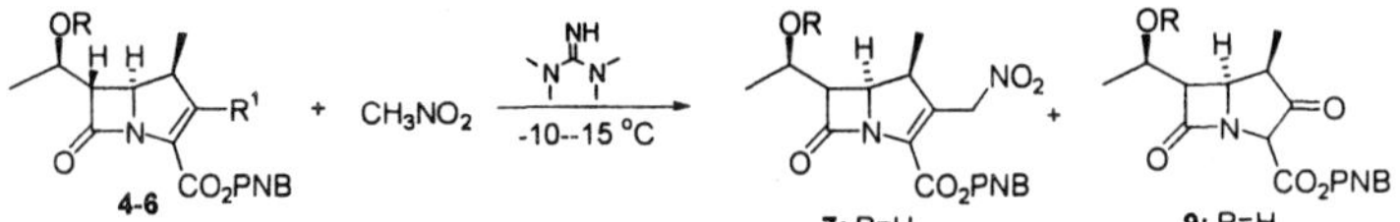

	SOLVENT	7: R=H / 8: R=TBS		9: R=H / 10: R=TBS
4: R=TBS, R¹=SOEt	CH_3NO_2	1	:	-
5: R=H, R¹=OPO(OPh)₂	CH_3NO_2	1	:	5
"	DMPU:CH_3NO_2 1:1	1	:	3
"	HMPA:CH_3NO_2 1:1	1	:	2.8
6: R=TBS, R¹=OTf	DMPU:CH_3NO_2 1:1	5-10	:	1

Figure 12. Nitromethane Displacements

Finally, it is my pleasure to formally acknowledge all of the superb co-workers who have made this program so successful. To the medicinal chemists, Milt Hammond, Ron Ratcliffe, Robert Wilkening and Frank DiNinno and their associates, I must convey my sincere admiration for a job superbly done. In preparation for this presentation, I had the chance to review their work in detail. I stand in awe of their ability to perform structure-activity studies on these structurally complex targets. I have already highlighted the key Process Research players and their individual contributions to this story. Their associates are noted in the references to the papers that have resulted from this work.[1d, 4, 5, 7, 10] As always, it is a great pleasure to have such talented, excellent and dedicated co-workers.

References

1. (a) Rosen, H.; Hajdu, R.; Silver, L. Kropp, H.; Dorso, K.; Kohler, J.; Sundello, J.G.; Huber, J.; Hammond, G.G.; Jackson, J.J.; Gill, C.J.; Thompson, R.; Pelak, B.A.; Epstein-Toney, J.H.; Lankas, G.; Wilkening, R.R.; Wildonger, K.J.; Blizzard, T.A.; DiNinno, F.P.; Ratcliffe, R.W.; Heck, J.V.; Kozarich, J.W.; Hammond, M.L. *Science* 1999, *283*, 703. (b) Wilkening, R.R.; Ratcliffe, R.W.; Wildonger, K.W.; Cama, L.D.; Dykstra, K.D.; DiNinno, F.P.; Blizzard, T.A.; Hammond, M.L.; Heck, J.V.; Dorso,

K.L.; St. Rose, E. Kohler, J.; Hammond, G.G. *Bioorganic and Med. Chem. Letters* **1999**, *9*, 673-678. (c) Ratcliffe, R.W.; Wilkening, R.W.; Wildonger, K.W.; Waddell, S.T.; Santorelli, G.M.; Parker, D.L. Jr.; Morgan, J.R.; Blizzard, T.A.; Hammond, M.L.; Heck, J.V.; Huber, J.; Kohler, J.; Dorso, K.L.; St. Rose, E.; Sundelof, J.G.; May, W.J.; Hammond, G.G. *Bioorganic and Med. Chem. Letters* **1999**, *9*, 679-682. (d) Yasuda, N; Huffman, M.A.; Jo, G.-J; Xavier, L.; Yang, C.; Emerson, E.M.; Tsay, F.-R.; Li Y.; Kress, M.H.; Rieger, D.L.; Karady, S.; Sohar, P.; Abramson, N.L.; DeCamp, A.E.; Mathre, D.J.; Douglas, A.W., Dolling, U.-H.; Grabowski, E.J.J.; Reider, P.J. *J. Org. Chem.* **1998**, *63*, 5438-5446.

2. Uyeo, S.; Itani, H. *Tetrahedron Lett.* **1994**, *35*, 4377-4381.

3. Kosugi, M.; Sumiay, T.; Ohhashi, K.; Sano, H.; Migita, T. *Chem. Lett.* **1985**, 997-998.

4. Yasuda, N.; Yang, C.; Wells, K.M., Jensen, M.S.; Hughes, D.L. *Tetrahedron Lett.* **1999**, *40*, 427-430.

5. Miller, R.A.; Humphrey, G.R.; Lieberman, D.R.; Ceglia, S.S.; Kennedy, D.J.; Grabowski, E.J.J.; Reider, P.J. *J. Org. Chem.* **2000**, *65*, 1399-1406.

6. Vedejs, E.; Haight, A.R.; Moss, W.O. *J. Am. Chem. Soc.* **1992**, *114*, 6556.

7. Jensen, M.S.; Yang, C.; Yi, H.; Rivera, N.; Wells, K.M.; Chung, J.L.; Yasuda, Y.; Hughes, D.L.; Reider, P.J. *Organic Lett.* **2000**, *2*, 1081-1084.

8. Reider, P.J.; Grabowski, E.J.J. *Tetrahedron Lett.* **1982**, *23*, 2293-2296.

9. Reider, P.J.; Grabowski, E.J.J. U.S. Patent 5,998,612 (1999).

10. Humphrey, G.R.; Miller, R.A.; Pye, P.J.; Rossen, K.; Reamer, R.A.; Maliakal, A.; Ceglia, S.S.; Grabowski, E.J.J.; Volante, R.P.; Reider, P.J. *J. Am. Chem. Soc.* **1999**, *121*, 11261-11266.

11. Chung, J.Y.L.; Grabowski, E.J.J.; Reider, P.J. *Organic Lett.* **1999**, *1*, 1783-1785.

Chapter 3

Synthesis of Steroid 'Antedrugs' for Evaluation as Safe Treatments for Asthma: Practical Synthesis of the Inhaled Steroid GR250495X

Matthew J. Sharp, Mark W. Andersen[1], Evan G. Boswell[2], Bobby N. Glover, Michael T. Martin, Richard T. Matsuoka, and Xiaoming Zhou

Chemical Development Department, Glaxo Wellcome Inc., 5 Moore Drive, Research Triangle Park, NC 27709 – 3398
[1]Current address: Magellan Laboratories, 160 Magellan Lab Court, Morrisville, NC 27560
[2]Current address: Schweizerhall, 1106 Perimeter Road, Donaldson Center, Greenville, SC 29605

The inhaled steroid antedrug candidate **GR250495X** was discovered at Glaxo Wellcome for the treatment of asthma. In order to fully evaluate the clinical potential of **GR250495X** an efficient synthesis of the drug candidate was required. Herein is described the development of an efficient and robust synthesis in which the key step is the coupling of 2(S)-mercapto-γ-butyrolactone (**19**) with the steroid acid (**15**).

Asthma is a chronic inflammatory condition of the airways. The most potent and effective agents available are the glucocorticoids. The introduction of the first inhaled glucocorticoid, beclomethasone dipropionate (BDP, **1**), in 1972 by Glaxo revolutionized asthma therapy (*1*). Delivery of the steroid topically to the lung dramatically reduced undesired systemic effects associated with oral steroid treatments. Even further reduction in systemic exposure was achieved with the introduction of fluticasone propionate (FP, **3**) in which hepatic inactivation affords the inactive carboxylic acid **4** as shown in Figure 1 (*2*). The search for even safer topical steroids has focused on plasma labile antedrugs (compounds designed to exert their effect locally but which are inactivated in the circulation to reduce unwanted systemic effects) and in particular the use of remote esterase labile functional groups (*3*). However, the ubiquitous distribution of esterase activity makes this approach less than ideal since premature hydrolysis in the lung is possible. The most striking example of this is the diester BDP (**1**) which is stable in blood plasma but which is rapidly converted to the more active monoester BMP (**2**) by esterases in the lung as shown in Figure 1 (*4*).

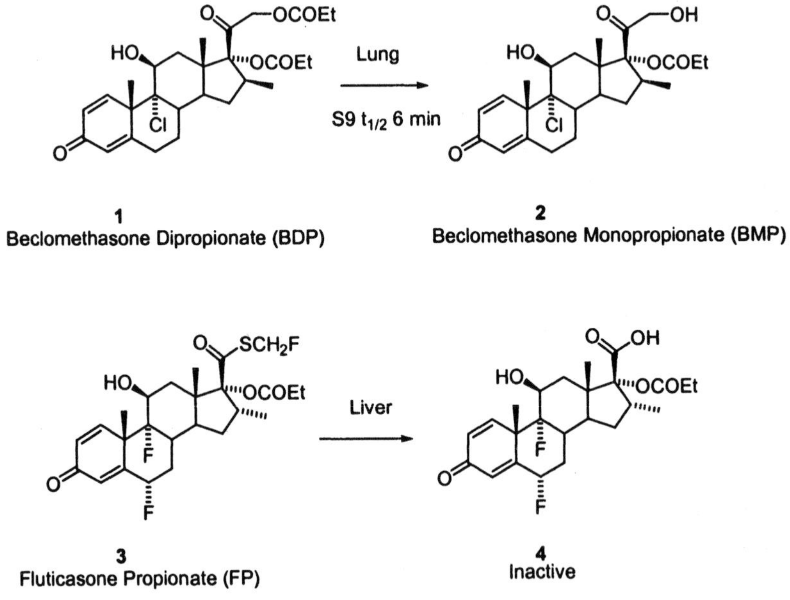

Figure 1. Beclomethasone Dipropionate and Fluticasone Propionate

More recently, the search for better antedrugs for the treatment of asthma at Glaxo Wellcome has focused on the incorporation of remote lactones onto the glucocorticoid nucleus (*5*). It was discovered that lactone hydrolysis is an ideal antedrug mechanism for inhaled steroids. For example, in the 21-thio derivatives of fluocinolone acetonide, the ethyl ester **5** showed some lability in blood plasma ($t_{1/2}$ 24 min) whereas; the corresponding lactone **6** was very rapidly hydrolyzed under the same conditions ($t_{1/2}$ <1min). Even more remarkably the lactones were found to be stable in human lung S9 preparations while the ethyl esters were rapidly cleaved under the same conditions (see Figure 2). Although, as early as 1966, it has been known that lactonase activity exists in human plasma, recent work at Glaxo Wellcome (5) suggests that this enzyme is human paraoxonase, an enzyme known to be absent from the lung, thus explaining the unique selectivity observed with this class of steroids.

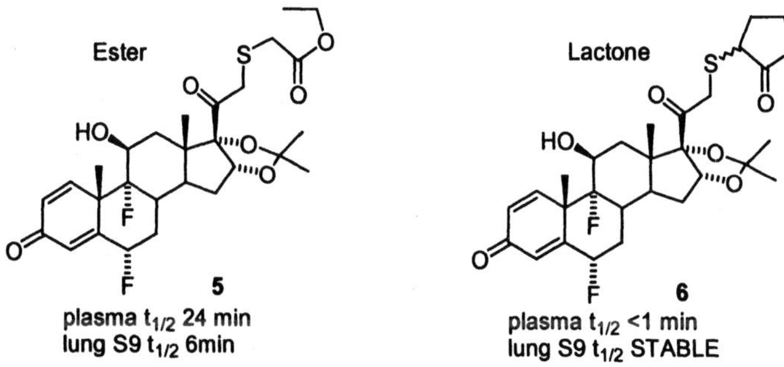

Ester

Lactone

5
plasma $t_{1/2}$ 24 min
lung S9 $t_{1/2}$ 6min

6
plasma $t_{1/2}$ <1 min
lung S9 $t_{1/2}$ STABLE

Figure 2. Comparison of lung and plasma stability of 21-thio fluocinolone acetonide derivatives containing remote ethyl ester and γ-butyrolactone groups

In this paper we describe the development of a practical synthetic route to the glucocorticoid antedrug **GR250495X**. This compound shows promise as an ideal antedrug candidate since it is stable in human lung S9, but is rapidly hydrolyzed to the corresponding inactive hydroxy acid **7** in blood plasma ($t_{1/2}$ 5 min) as shown in Figure 3.

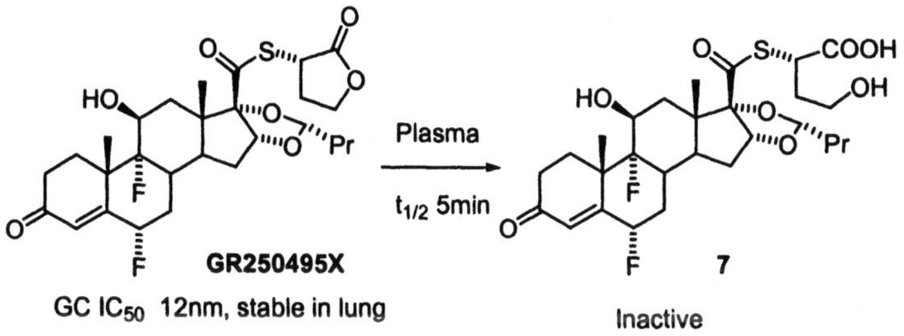

Plasma
$t_{1/2}$ 5min

GR250495X
GC IC$_{50}$ 12nm, stable in lung

7
Inactive

Figure 3. Inactivation of GR250495X in blood plasma

Synthesis of GR250495X

The most convenient starting material for the synthesis of **GR250495X** is the commercially available fluocinolone acetonide (**8**). To convert fluocinolone acetonide to **GR250495X** the following synthetic transformations must be accomplished: the selective reduction of the $\Delta^{1,2}$ olefin, oxidation at C-20, stereoselective incorporation of (S)-2-mercapto-γ-butyrolactone, and selective conversion of the 16,17-acetonide to the 16,17 (R-butylidenedioxy) group as shown in Figure 4. Another key synthetic challenge in the synthesis of **GR250495X** is the development of a proprietary route of manufacture. Given the intense competition in the anti-inflammatory steroid arena, avoiding competitors' intellectual property is a nontrivial challenge. One more factor that must be considered in the synthesis is that **GR250495X** contains a readily epimerizable stereocenter alpha to the lactone carbonyl. Careful consideration must therefore be given to the timing of incorporation of this stereocenter, and the subsequent reaction conditions.

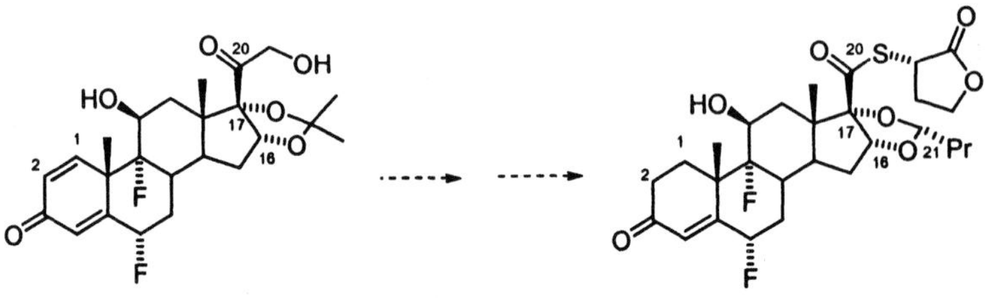

8
Fluocinolone Acetonide

GR250495X

Figure 4. Fluocinolone acetonide as a starting material in the synthesis of GR250495X

Route 1

The first synthetic route used to prepare **GR250495X** starts with an oxidation at C-20 as shown in Figure 5. Fluocinolone acetonide (**8**) was converted to the corresponding C-17 acid **9** by the K_2CO_3 catalyzed air oxidation developed by Kertesz and Marx (*6*). This transformation was convenient on small scale because the reaction mixture may be simply exposed to the atmosphere. But on larger scale (>1 liter) air or oxygen must be passed through the mixture to maintain the reaction, which leads to evaporation of the solvent and potential safety concerns (a flammable solvent in the presence of oxygen) in future manufacturing. Acidification of the reaction mixture afforded the crystalline acid **9** which was isolated by filtration in 80% yield. Two methods were investigated for the selective reduction of the $\Delta^{1,2}$ olefin: homogeneous hydrogenation with Wilkinson's catalyst as originally reported by Djerassi (*7*), and a more recently reported copper catalyzed hydrogenation (*8*).

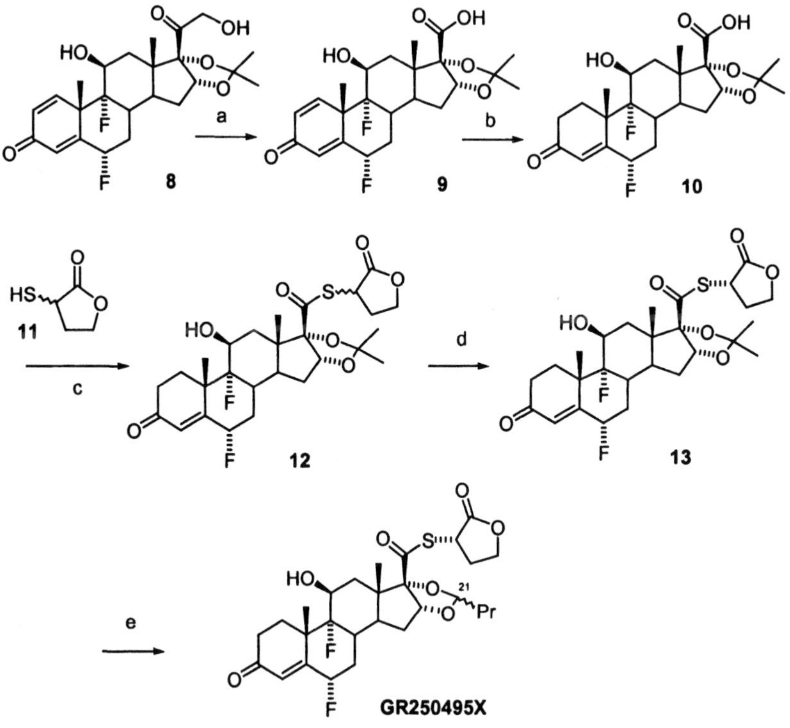

Reagents and conditions: (a) air, EtOH, K_2CO_3, 80%; (b) Rh(PPh$_3$)$_3$Cl, H$_2$, EtOH-toluene, 56 °C, 65%; (c) HBTU, NEt$_3$, DMF, 80%; (d) NEt$_3$, methyl isobutyl ketone, 90%; (e) PrCHO, 70% H$_2$SO$_4$, CH$_2$Cl$_2$, 85%.

Figure 5. Route 1 synthesis of GR250495X

Of the two methods, homogeneous hydrogenation in the presence of Wilkinson's catalyst gave superior results and provided the 1,2-dihydro steroid in reasonable yield. However, problems were again encountered on scaling this reaction. Due to the low solubility of the substrate and product in the reaction solvent (EtOH-toluene), high dilutions, long reaction times (7 to 15 days), and multiple charges of fresh catalyst were needed to drive the selective hydrogenation to completion. It was found necessary to completely consume all starting material since any remaining diene is difficult to remove, and is carried on to drug substance. Basic extraction followed by acidification afforded the crystalline 1,2-dihydro acid **10** in 65% yield. The selective incorporation of the (S)-2-mercapto-γ-butyrolactone was accomplished in two steps. First the HBTU activated 1,2-dihydro acid **10** was coupled with racemic 2-mercapto-γ-butyrolactone (**11**) (*9*) in the presence of triethylamine to afford the 1:1 mixture of lactone diastereomers **12**, that were isolated by crystallization in 80% yield. This diastereomeric mixture was then suspended in methyl isobutyl ketone containing 0.5 eq of triethylamine and stirred at 32 °C for 48 hours. Filtration of the resulting mixture afforded the desired lactone diastereomer **13** in 90% yield that contained less than 0.7% of the undesired diastereomer by HPLC analysis. This remarkable increase in diastereomeric ratio has been described in related systems as a crystallization induced second-order asymmetric transformation (*10*). The final transformation in this route was the conversion of 16α,17α-acetonide to the corresponding 16α,17α-butylidenedioxy group. With acetal formation it is necessary to control the stereochemistry of the newly formed C-21 center. It is known that in similar steroidal 16α,17α acetals the desired 21(R) diastereomer is thermodynamically preferred (*11*). Fortunately, it was discovered that treatment of the acetonide **13** with butyraldehyde in the presence of catalytic H_2SO_4 afforded the desired 21(R)-acetal **GR250495X** as the major diastereomer (10:1 C-21 R/S ratio) in 85% yield. However, removal of the undesired minor 21(S)-acetal diastereomer by crystallization proved to be very inefficient. For example, three sequential recrystallizations of the crude product from MIBK only marginally improved the diastereomeric purity of GR250495X (20:1 C-21 R/S ratio) in only 40% recovery. Although the final stage of Route 1 was quite inefficient (providing **GR250495X** contaminated with 5% of the undesired 21(S)-acetal) this route had the advantage of possessing no intellectual property issues, and was therefore used to prepare the initial quantities of drug substance for pre-clinical and clinical evaluation (ca. 400g).

Route 1a

In order to support the further development of the drug candidate **GR250495X**, a more efficient synthesis was needed that was amenable to scale-up and long term manufacturing. Initially, we explored switching the order of the synthetic stages as shown in Figure 6. It was found to be advantageous to start the synthesis not with the C-20 oxidation, but with the selective hydrogenation of the $\Delta^{1,2}$ olefin. Carrying out this reduction on the more soluble C-20 ketone analogue allowed for a higher throughput of material with shorter reaction times and a lower catalyst loading. Thus atmospheric hydrogenation of fluocinolone acetonide (**8**) in the presence of 5 mole % of Wilkinson's catalyst in EtOH at 50 °C resulted in complete consumption of starting material in 48hrs.

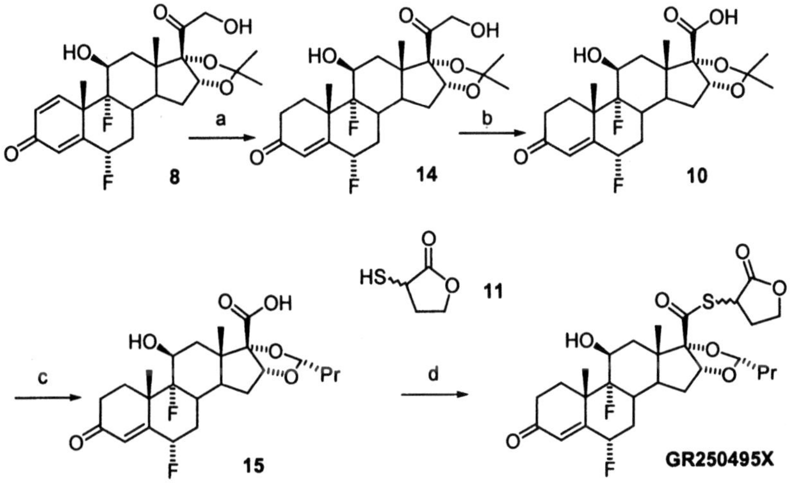

Reagents and conditions: (a) 5% Rh(PPh$_3$)$_3$Cl, H$_2$, EtOH, 50 °C, 88%; (b) H$_2$O$_2$, K$_2$CO$_3$, MeOH, 80%; (c) i. PrCHO, HClO$_4$, CH$_2$Cl$_2$, -5 °C, ii. EtOH-H$_2$O recryst, 65%; (d) HBTU, NEt$_3$, DMF, 80%.

Figure 6. Route 1a synthesis of GR250495X

Crystallization of the product by the addition of water afforded 1,2-dihydrofluocinolone acetonide (**14**) in 88% yield containing less than 0.5% of the starting diene by HPLC analysis. Another advantage of starting the synthesis with the hydrogenation of fluocinolone acetonide is that we expect the 1,2-dihydrofluocinolone acetonide to become commercially available in the future, thus saving a synthetic stage. Typically the corticosteroids are manufactured by partial synthesis from stigmasterol or sitosterol extracted from soybeans, or from diosgenin isolated from the Mexican yam and in all cases the $\Delta^{1,2}$ olefin must be introduced synthetically (*12,13*).

Next we turned our attention to the oxidation at C-20. Interestingly, the air oxidation used in Route 1 was found to be less efficient when attempted on 1,2-dihydrofluocinolone acetonide. Longer reaction times were needed and a significant side reaction was oxidation at C-6 to give the corresponding 6-keto derivative. Fortunately, however, a more practical method was found to accomplish this oxidation. Treatment of 1,2-dihydrofluocinolone acetonide (14) with aqueous H_2O_2 and K_2CO_3 in MeOH (14) at room temperature cleanly afforded the 1,2-dihydro acid 10 in 80% yield after extractive work-up followed by crystallization. This reaction was found to be more amenable to scale-up and did not have the safety liabilities of the air oxidation. The next challenge in the sequence was the conversion of the 16α,17α-acetonide 10 to the corresponding 16α,17α-butylidenedioxy derivative 15. We were encouraged to find that acetalization of 10 with butyraldehyde in the presence of catalytic perchloric acid in dichloromethane at -5 °C gave the desired 21(R)-acetal 15 as the major diastereomer (10:1 R/S ratio). More importantly, recrystallization of the crude product from EtOH-H_2O removed the undesired diastereomer, affording the desired 21(R)-acetal (15, >160:1 R/S ratio) in 65% overall yield. Thus, we now had an efficient synthetic route to an advanced key intermediate containing the desired 16α,17α-(R)-butylidenedioxy functionality. Although there were some short term intellectual property issues associated with the use of the key intermediate 15, these would not effect longer term manufacturing of the product because the relevant patents are expected to expire before possible commercialization of our drug candidate. All that remained in the synthesis was the stereoselective introduction of the 2(S)-mercapto-γ-butyrolactone group. However, while coupling the acetal-acid 15 with racemic 2-mercapto-γ-butyrolactone (11) afforded the 1:1 mixture of diastereomers in good yield, all attempts to convert this mixture to pure GR250495X by a selective crystallization-induced transformation failed. Unlike in the acetonide example above, the two lactone diastereomers of the butylidene series were found to have similar solubility in all suitable solvents, thus rendering a crystallization approach ineffective. We therefore turned our attention to the enantioselective synthesis of 2(S)-mercapto-γ-butyrolactone, since coupling of this with our key intermediate 15 should provide pure GR250495X.

Route 2

The successful synthesis of the novel 2(S)-mercapto-γ-butyrolactone (19) is shown in Figure 7. This synthesis starts with 2(R)-hydroxy-γ-butyrolactone (16) which is commercially available or may be conveniently prepared from d-malic acid (15). Conversion of 2(R)-hydroxy-γ-butyrolactone (16) to the corresponding crystalline 2(R)-mesylate 17, was accomplished by treatment with methanesulfonyl chloride in the presence of DMAP in EtOAc. The mesylate could be isolated by crystallization from the reaction mixture after work-up. Fortunately, a remarkable enantiomeric purification occurred during this crystallization and mesylate with >99.8%ee could be obtained from starting alcohol of only 90%ee.

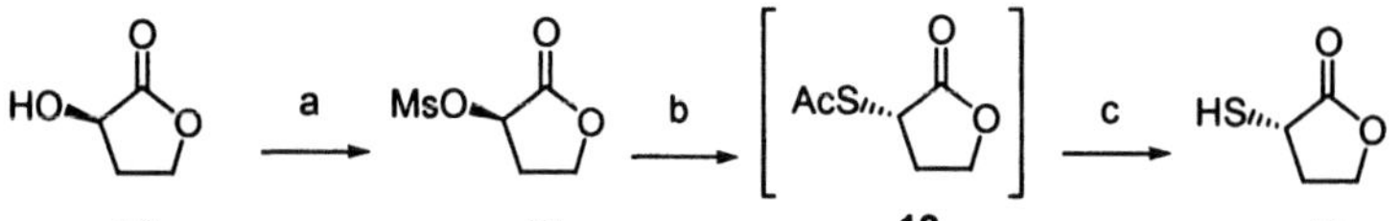

16 90%ee **17** >99.8%ee **18** **19** >97%ee

Reagents and conditions: (a) i. MsCl, DMAP, NEt_3, EtOAc, 10-20 °C, ii. EtOAc recryst. 64%; (b) AcSK, toluene; (d) i. HCl, MeOH-toluene, ii. distillation, 70%.

*Figure 7. Synthesis of 2(S)-mercapto-γ-butyrolactone (**19**)*

Conversion of the mesylate **17** to 2(S)-mercapto-γ-butyrolactone (**19**) was accomplished in a two step process. Displacement of (R)-mesylate with potassium thioacetate in toluene cleanly afforded the (S)-thioacetate **18,** which was not isolated, but cleaved to the thiol by treatment with MeOH and anhydrous HCl. Distillation then afforded 2(S)-mercapto-γ-butyrolactone (**19**) in 70% overall yield and >97%ee.

An alternative enzymatic synthesis of 2(S)-mercapto-γ-butyrolactone (**19**) was also attempted as shown in Figure 8. Surprisingly, there are relatively few examples of enzymatic resolution of thiols, however, we were encouraged by the work of Cesti (*16*) to investigate the lipase mediated hydrolysis of racemic thioacetate **20** (*9*).

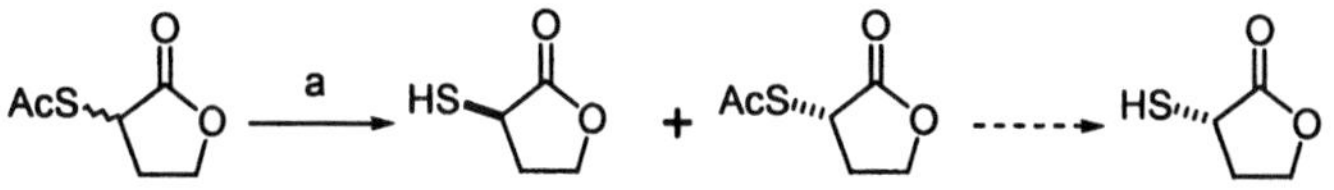

20 **21** (1.5 : 1) **18** 90%ee **19**

Reagents and conditions: (a) *Mucor meihei* lipase, nBuOH, toluene-H_2O

*Figure 8. Enzymatic synthesis of 2(S)-mercapto-γ-butyrolactone (**19**)*

For this approach to be successful we would not only need a high degree of stereoselectivity in the hydrolysis, but also a high degree of chemoselectivity, with preferential hydrolysis of the thioacetate over the lactone. To test this, racemic thioacetate **20** was treated with several commercially available lipases in both aqueous and organic systems, and reaction aliquots removed and analyzed by chiral GC. It was found that treatment of **20** with *Mucor Meihei* lipase in the presence of nBuOH in a toluene-H_2O mixture resulted in a selective trans-thioesterification. Moreover, this enzyme preferentially reacted with the *R* enantiomer, leaving the reaction mixture enriched in the (S)-thioacetate **18** and (R)-thiol **21**. Unfortunately, even under optimum conditions, the yield and selectivity of this enzymatic process were not competitive with the two-stage process described above.

With a source of 2(S)-mercapto-γ-butyrolactone (**19**) in hand we next turned our attention to coupling this molecule with the key steroid intermediate **15** as shown in Figure 9. If this route was to be successful, we needed to find a mild method to form the thioester bond without racemization of 2(S)-mercapto-γ-butyrolactone, or epimerization of the **GR250495X** product. It soon became apparent that standard coupling conditions involving tertiary amine bases could not be utilized in this reaction since thiol and reaction product are rapidly racemized and epimerized, respectively, by even the most hindered of these bases.

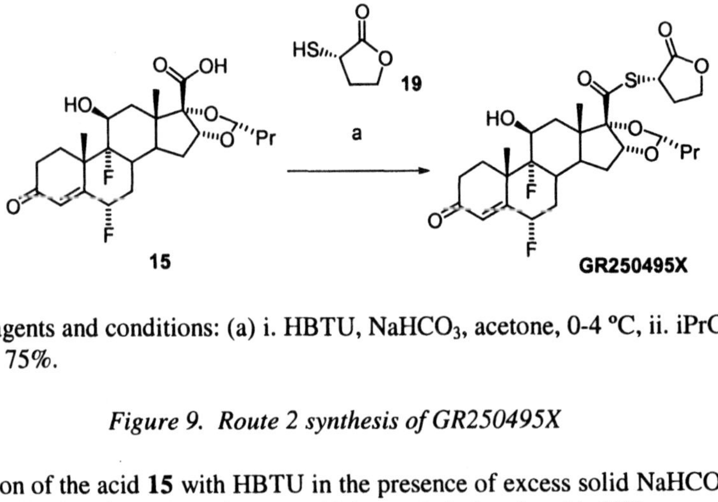

Reagents and conditions: (a) i. HBTU, $NaHCO_3$, acetone, 0-4 °C, ii. iPrOAc recryst., 75%.

Figure 9. Route 2 synthesis of GR250495X

Activation of the acid **15** with HBTU in the presence of excess solid $NaHCO_3$ in acetone afforded the activated ester, which coupled cleanly with 2(S)-mercapto-γ-butyrolactone at 0-4 °C with minimal detectable epimerization during the reaction. Crystallization of the crude product from the reaction mixture afforded **GR250495X** in 75% yield containing less than 2% of the undesired lactone diastereomer. An additional recrystallization from iPrOAc-heptane provided the drug candidate **GR250495X** in 72% recovery and lowered the level of this impurity to below 0.6%.

Route 2 was found to have several advantages over Route 1. Firstly, Route 2 gave higher quality drug substance with the major impurity being 0.6% of the undesired *R*-lactone diastereomer, as opposed to 5% of the undesired *S*-acetal diastereomer from Route 1. Secondly, Route 2 was found to be more efficient, requiring only four linear stages from 1,2-dihydrofluocinolone acetonide in 28% overall yield compared with six linear stages from fluocinolone acetonide in 8% overall yield for Route 1. Route 2 was also found to be amenable to scale-up allowing the rapid synthesis of 400g of **GR250495X** from 20L glassware. This route was also free of any long-term intellectual property issues, and was therefore our preferred route for future manufacturing. Although the utilization of Route 2 for the synthesis of **GR250495X** seemed an obvious choice, there existed a third possibility that offered the potential for an even more direct synthesis.

Route 3

A potentially more efficient synthesis of **GR250495X** is shown in Figure 10. Since we now had a practical route to the crystalline *R*-mesylate **17** (>99.8%ee), and we had shown that coupling of this fragment with thioacetic acid proceeded with clean inversion, we decided to investigate coupling directly with the thioacid **22** derived from the key intermediate **15**. The synthesis of the thioacid **22** was most conveniently carried out via the initial activation of the carboxylic acid moiety of acid **15** with 1,1'-carbonyldiimidazole (CDI) in DMF followed by the treatment of the intermediate acyl imidazolide with hydrogen sulfide gas. Subsequent acidification of the reaction mixture with aqueous hydrochloric acid solution induced precipitation of the desired thioacid **22** which could be carried directly into the next step without further purification. The next and final step of Route 3 involved thioester formation with *R*-mesylate **17**. The challenge in this step was to find reaction conditions, which would afford ester formation with clean inversion of configuration at the stereogenic center of the lactone and without oxidation of the thio acid. Thio acid **22** must be isolated, if made using CDI, before treatment with *R*-mesylate **17**. If thioacid **22** is not isolated, the presence of residual imidazole originating from CDI during the esterification leads to complete racemization of the lactone's stereogenic center in **GR250495X**. Initial conditions using potassium carbonate in toluene yielded the desired product with approximately 98% d.e. Unfortunately, the rate of this esterification in toluene was extremely slow, taking over 2 days to go to completion at room temperature.

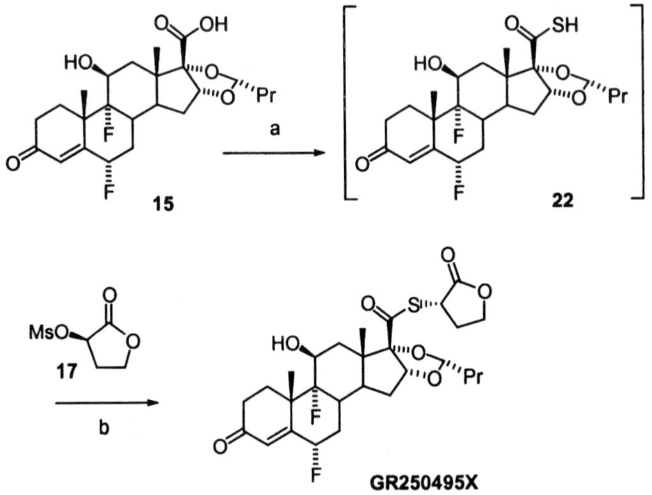

Reagents and conditions: (a) i. CDI, DMF, ii. H₂S; (b) CsCO₃, DMF.

Figure 10. Route 3 synthesis of GR250495X

After screening a number of solvents and both soluble and insoluble bases, we found that the use of cesium carbonate in DMF at room temperature afforded **GR250495X** with 99.4% d.e. in less than 1 h. Subsequent acidification of the reaction mixture with dilute aqueous hydrochloric acid solution induced precipitation of **GR250495X** in greater than 70% yield over the two steps. The final product was purified via a recrystallization (isopropyl acetate / heptane) which afforded the target molecule with a slightly enhanced 99.7% d.e.

Although Route 3 is conceptually elegant, providing **GR250495X** in only one stage from the key intermediate **15** in 70% overall yield and 99.7% d.e., two new impurities were introduced by this process. The first impurity is the *R*-mesylate **17**, found to be present in **GR250495X** synthesized via this route at levels in excess of 2000 ppm even after recrystallization. The second impurity is the disulfide dimer presumably formed via the oxidative coupling of the thioacid **22**. This impurity was typically present in the recrystallized final product at levels over 3%. Carrying out the thioacid preparation in the presence of dithiothreitol, a reagent for the cleavage of disulfide bonds (*17,18*), greatly minimizes the formation of this unwanted disulfide dimer. Unfortunately, dithiothreitol is relatively expensive and difficult to remove from thioacid **22**. Efforts are currently ongoing to optimize Route 3 such that it will provide drug substance free of the above two impurities. This route has the advantage of being fewer stages than Route 2 and does not require the preparation or distillation of the sensitive 2(S)-mercapto-γ-butyrolactone (**19**).

Summary

Four synthetic routes to the inhaled glucocorticoid antedrug candidate, **GR250495X**, were evaluated. Two of these routes were utilized to prepare significant quantities of the drug candidate. In Route 1 **GR250495X** was prepared in six stages and 8% overall yield from fluocinolone acetonide and the major impurity was the *S*-acetal derivative (5% area by HPLC). The key step in this route was a crystallization-induced asymmetric transformation. The major advantage of this route was that it was free of any intellectual property issues that could delay or limit the development of the drug candidate. Using this route, 400 grams of **GR250495X** were rapidly prepared for use in pre-clinical and early clinical evaluation. Route 2 was found to be more efficient consisting of four synthetic stages from dihydrofluocinolone acetonide, and **GR250495X** was prepared in 28% overall yield with the major impurity being the *R*-lactone derivative (0.5% area by HPLC). The key step in Route 2 was the coupling of 2(S)-mercapto-γ-butyrolactone with the key steroid intermediate **15**. This route was also more amenable to scale-up and is currently the preferred process for future synthesis of the drug candidate. Route 3 has the potential of being the most efficient route to **GR250495X**. The key step is the coupling of steroid thioacid **22** with the *R*-mesylate **17**. Work is ongoing to optimize this process.

Acknowledgements

We gratefully acknowledge Purnima Narang, Amita Thakrar and Nik Chetwyn for providing analytical support for this project, and Keith Biggadike, Clive Meerholz and John Partridge for contributing in many helpful discussions.

References

1. Barnes, P. J.; Pedersen, S.; Busse, W. W. *Am. J. Respir. Crit. Care Med.* **1998**, *157*, S1.
2. Hogger, P.; Rohdewald, P.; McConnell, W.; Howarth, P.; Meibohm, B.; Mollmann, H.; Wagner, M.; Hochhaus, G.; Mollmann, A.; Derendorf, H.; Neil Johnson, F.; Barnes, N. C.; Hughes, J.; Williams, J.; Pearson, M. G. *Rev. Contemp. Pharmacother.* **1988**, *9*, 501.
3. Kwon, T.; Heiman, A. S.; Oriaku, E. T.; Yoon, K.; Lee, H. J. *J. Med. Chem.* **1995**, *38*, 1048 and references therein.
4. Wurthwien, G.; Rohdewald, P. *Biopharm. Drug Dispos.* **1990**, *11*, 381.

5. Biggadike, K.; Angell, R. M.; Burgess, C. M.; Farrell, R. M.; Hancock, A. P.; Harker, A. J.; Irving, W. R.; Ioannou, C.; Procopiou, P. A.; Shaw, R. E.; Solanke, Y. E.; Singh, O. M. P.; Snowden, M. A.; Stubbs, R. J.; Walton, S.; Weston, H. E. *J. Med. Chem.* **2000**, *43*, 19.

6. Kertesz, D. J.; Marx, M. *J. Org. Chem.* **1986**, *51*, 2315.

7. Djerassi, C.; Gutzwiller, J. *J. Am. Chem. Soc.* **1966**, *88*, 4537.

8. Ravasio, N.; Rossi, M. *J. Org. Chem.* **1991**, *56*, 4329.

9. Fuchs, G. *Ark. Kemi.* **1966**, *26*, 111.

10. Jaques, J.; Collet, A.; Willen, S. H. Enantiomers, Racemates and Resolutions; Wiley: New York, 1981 and references cited therein.

11. Ashton, M. J.; Lawrence, C.; Karlsson, J-A.; Stuttle, K. A. J.; Newton, C. G.; Vacher, B. Y. J.; Webber, S.; Withnall, M. J. *J. Med. Chem.* **1996**, *39*, 4888.

12. Hogg, J. A. *Steroids*, **1992**, *57*, 593.

13. Lenz, G. R. *In Kirk-Othmer Encyl. Chem. Technol.*, 3rd Ed. **1983**, *21*, 645.

14. Nguyen, T. T.; Kringstad, R.; Aasen, A. J.; Rasmussen, Knut, E. *Acta Chem. Scand., Ser. B*, **1988**, *B42*, 403.

15. Shiuey, S-J.; Partridge, J. J.; Uskokovic, M. R. *J. Org. Chem.* **1988**, *53*, 1040.

16. Bianchi, D.; Cesti, P. *J. Org. Chem.* **1990**, *55*, 5657.

17. Kumar, P.; Bose, N. K.; Gupta, K. C. *Tetrahedron Lett.* **1991**, *32*, 967.

18. Hovinen, J.; Guzaev, A.; Azhayev, A.; Lonnberg, H. *Tetrahedron Lett.* **1993**, *34*, 8169.

Chapter 4

Development of a Catalytic, Asymmetric Michael Addition in the Synthesis of Endothelin Antagonist ABT-546

David M. Barnes, Jianguo Ji, Ji Zhang, Steven A. King,
Steven J. Wittenberger, and Howard E. Morton

Process Chemistry Research, Pharmaceutical Products Division,
Abbott Laboratories, Building R8/1, 1401 Sheridan Road,
North Chicago, IL 60064

The development of a highly enantioselective catalytic asymmetric conjugate addition of 1,3-dicarbonyl compounds to nitroalkenes is described, as is its use in the synthesis of the selective endothelin A antagonist ABT-546. Employing 4 mol% of a bis(oxazoline)-Mg(OTf)$_2$ complex with 5.5 mol% of an amine co-catalyst, the product nitroketone is obtained in 88% ee, with good yields. Particularly important to the reaction is the water content. While water is necessary during the generation of the catalyst, the water must then be removed in order to maximize selectivity and reactivity. The scope of the reaction has been extended to other ketoester and malonate nucleophiles, as well as other nitroolefins.

Endothelin-1 (ET) is a small (21 amino acid) peptide which is the most potent known vasoconstrictor. Its activity is mediated *via* its interaction with two receptors, endothelin A (ET_A) and endothelin B (ET_B). While ET_B is a shunt receptor which consumes ET, ET_A is implicated in vasoconstriction as well as proliferative effects. Therefore, selective antagonists of the ET_A receptor have been targeted as possible therapeutic agents for the treatment of cancer and congestive heart failure (1,2).

Shown in Figure 1 are two Abbott compounds which have been selected for clinical development (3). Both are tetrasubstituted pyrrolidines, with three contiguous stereocenters arrayed in the thermodynamically favored trans-trans configuration. ABT-627, with an aryl substituent at C-2, is currently in Phase II clinical trials. More recently ABT-546, with an aliphatic substituent at C-2, was identified as a potential backup. While it is less potent than ABT-627, ABT-546 is significantly more selective, suggesting that it may possess a superior therapeutic window. When ABT-546 was selected as a clinical backup to ABT-627, it was estimated that 3 Kg of material would be required for initial toxicology and Phase I clinical trials.

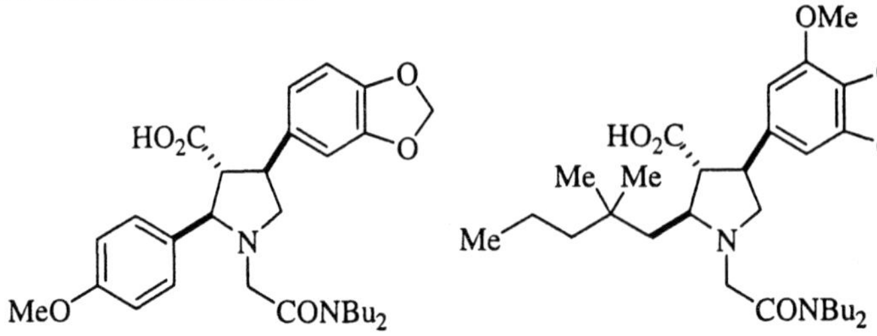

	ABT-627	ABT-546
Potency ET-A Ki (nm)[5]	0.034	0.46
ET-A selectivity (vs ET-B)[5]	1900	28000

Figure 1. Selective pyrrolidine inhibitors of Endothelin A

Early Synthetic Work

A number of routes can be envisioned for the preparation of polysubstituted pyrrolidines, and the intramolecular addition of a silyl ketene acetal to an oxime ether has been employed in the synthesis of ABT-627 (4). However, the reductive cyclization of a nitroketone, which could be obtained via the conjugate addition of a ketoester to a nitrostyrene, was viewed as being the most direct, convergent approach to ABT-546. In this approach, the nitro functionality is reduced to generate the cyclic imine, which is then further reduced to the trisubstituted pyrrolidine (Figure 2). As the *trans-trans* isomer is thermodynamically favored, both it and the *cis-cis* isomer are usable intermediates.

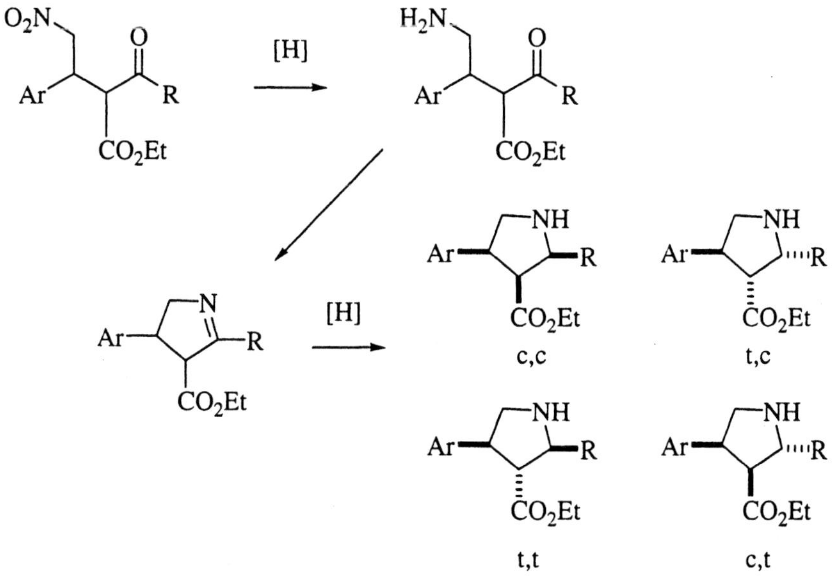

Figure 2. The nitroketone approach

In fact, as shown in Figure 3, this approach has been employed in the synthesis of ABT-627. The coupling of ketoester **1** with nitrostyrene **2** was accomplished with a catalytic quantity of KOtAm. Product nitroketone **3** was reduced over Raney nickel to prepare cyclic imine **4**, which was further reduced on addition of TFA to produce the *cis-cis* pyrrolidine **5** with good selectivity. Following epimerization with DBU, trans-trans pyrrolidine **6** was resolved with good efficiency as the mandelate salt.

With this encouraging precedent, a similar approach was investigated in the first-generation synthesis of ABT-546 (Figure 4). The Michael addition was again carried out in the presence of catalytic base. The reductive cyclization of nitroketone **9** was carried out employing modified conditions to yield directly the *trans-trans* pyrrolidine, **10**. However, despite significant effort, only tartaric acid was found to resolve this pyrrolidine, and then the optically enriched product was obtained with low recovery.

It was noted that the C-2 and C-3 stereocenters of pyrrolidine **10** were derived from the aryl-bearing stereocenter of the nitroketone. Therefore we considered methods whereby we might obtain enantioselectively the nitroketone

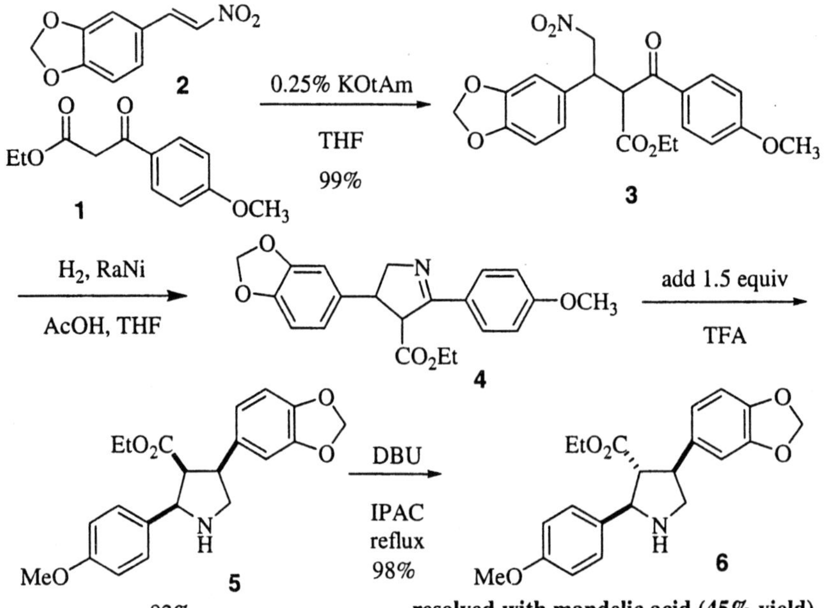

resolved with mandelic acid (45% yield)

Figure 3. Preparation of ABT-627

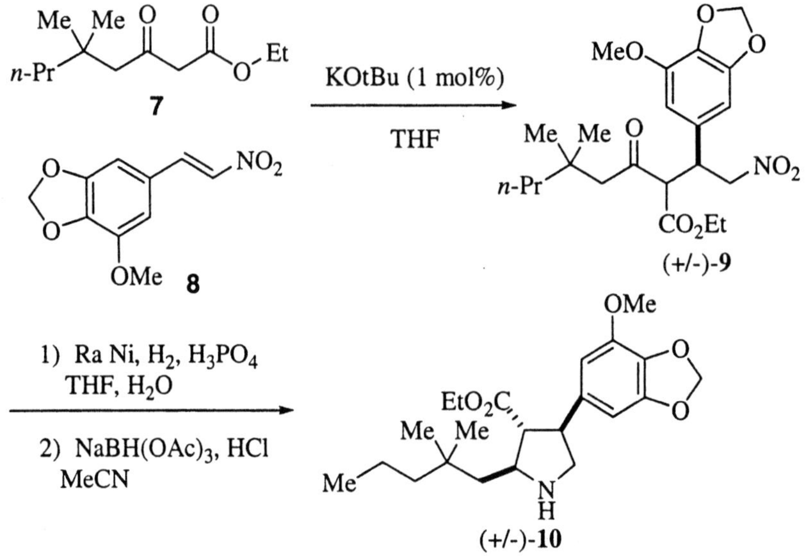

resolved with tartaric acid (20% yield)

Figure 4. First generation synthesis of ABT-546

or a similar synthon. A number of disconnections were entertained in which two achiral partners could be joined *via* the mediation of a chiral catalyst or auxiliary, though none were well-precedented in the literature. We were again attracted to the ketoester/nitrostyrene Michael disconnection, due to its convergency, and because some related Michael additions have been described. (*5*)

Development of the Michael Addition

Not many highly enantioselective catalytic additions to nitroalkenes have been reported (*6*), and none involving the use of β-ketoesters as the nucleophile (*7*). A number of metal/ligand combinations were screened, as were chiral auxiliaries. However, nothing provided significant selectivities (>20% ee) until the two reacting partners were combined in the presence bis(oxazoline) **11** and Mg(OTf)$_2$ in ethanol-stabilized CHCl$_3$ at elevated temperatures (eq 1). Then, the product nitroketone was obtained with moderate though variable selectivities (*8*). Other metal salts were ineffective in promoting the reaction selectively.

$$(1)$$

Optimization

It was not clear at first whether the catalyst was acting by coordinating to the nitro functionality, activating the nitrostyrene to attack by the enol tautomer of the ketoester. Alternatively, it could activate the ketoester towards addition to the nitrostyrene. An initial indication that the latter hypothesis was correct was obtained by the addition of 1 equiv. of an amine base as cocatalysts (Figure 5). Under these conditions (20 mol% catalyst), the reaction was run at ambient temperature, with selectivities approaching 80% ee. In fact, it was found later that the ketoester contained 1-2 mol% of imidazole, which accounted for the initial reaction. There was a relative lack of effect of the amine structure on the reaction, though stronger bases such as Et₃N delivered lower selectivities due to a background rate. It was further found that 1 equiv. of amine (N-methylmorpholine, NMM) relative to catalyst was sufficient to provide full activity. These results supported a soft enolization mechanism in which the catalyst-bound ketoester is deprotonated by the base, generating a magnesium enolate with chirality on the ligand which is transferred by diastereoselective addition to the nitrostyrene (eq 2).

Amine	ee	NMM equiv. (rel. to metal)	Conv. (5h)
2,6-lutidine	77%		
imidazole	74%	0	4% (24h)
5,6-dimethylbenzimidazole	80%	0.5	33%
N-methylmorpholine (NMM)	70%	1	46%
N-ethylpiperidine	64%	2	49%
Et₃N	68%	5	46%

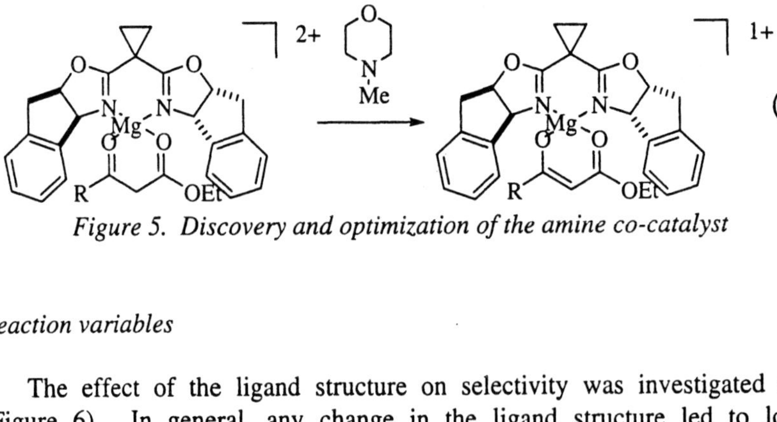

(2)

Figure 5. Discovery and optimization of the amine co-catalyst

Reaction variables

The effect of the ligand structure on selectivity was investigated next (Figure 6). In general, any change in the ligand structure led to lower selectivity, and usually to lower activity. For example, replacing the cyclopropane bridge with dimethyl (11 -> 12) led to a dramatic decrease in selectivity, as well as a 10-fold decrease in reactivity. Only cyclopropane-bridged diphenyl ligand 15 provided comparable selectivities.

The effect of other reaction variables was also investigated. The triflate counterion was optimal (80% ee), with coordinating counterions leading to lower reaction selectivity (I; 65% ee; Br; 22% ee). Other solvents were

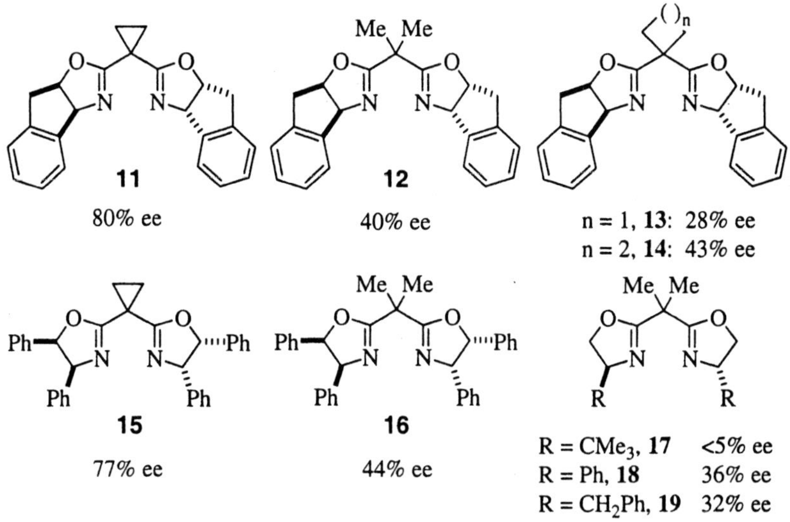

Figure 6. Effect of ligand structure on selectivity

investigated, including toluene (80% selectivity), EtOAc (60% selectivity), and 1,2-DCE (46% selectivity); chloroform and toluene have proven optimal for this reaction. However, due to the low solubility of nitrostyrene in toluene, those reactions must be run at elevated temperatures to achieve reasonable reaction rates. In particular, we were intrigued by the difference between chloroform (ethanol-stabilized, 80% ee) and CH_2Cl_2 (65% selectivity), feeling that the ethanol stabilizer might contribute to the difference. However, the use of hydrocarbon-stabilized chloroform increased the selectivity to 87%, and the reaction proceeded about 20% faster!

Water

As hydroxylic contaminants interfere with the reaction, we next investigated the effect of water on the reaction. When run at ambient temperature with 10 mol% catalyst, the reaction proceeded to 17% conversion in 25 h, with 86% selectivity. However, when 4A molecular sieves were employed, 55% conversion and 91% selectivity was obtained in the same time. From a practical standpoint, these results allowed the reaction to be run with only 4 mol% of the catalyst, with mild heating. Control experiments showed that the sieves served only to remove water from the reaction medium.

KF coulometry indicated that the commercially available $Mg(OTf)_2$ had been obtained as the tetrahydrate, and that the salt accounted for 80% of the water introduced into the reaction. This hydration was critical to catalyst generation. When the catalyst was generated from tetrahydrate, the reaction (4 mol% catalyst, 5.5 mol% N-methylmorpholine at 35 °C) proceeded to 38% conv. in 3.5 h. However, when the anhydrous (0.2 equiv. water by KF coulometry) Mg(OTf)$_2$ was employed, only 16% conv. was observed in the same time, though the reaction selectivity was unchanged. The activity was fully regenerated by adding 4 equiv. of water to the solid Mg(OTf)$_2$.

Practically, the reaction is carried out by first hydrating the salt, then combining it with the ligand in CHCl$_3$. After stirring an hour, the ketoester and nitrostyrene are added along with the molecular sieves. After the reaction is judged to be dry by KF coulometry, NMM is added to initiate the reaction.

The basis for this water requirement is not understood at the present time. Empirically, the reaction behavior (the lack of effect on selectivity) suggests that in the absence of water the catalyst is not fully formed, or that an inactive species is formed instead. However, the nature of that species is not currently known.

Further optimization

When reactions were run in order to optimize the concentration of the reaction, we found that the reaction rate did not vary. Practically this meant that other concerns could determine reaction concentration (0.2 M was generally found to be most convenient, and to provide slightly higher selectivities). But it also suggested that the reaction was first order in catalyst, and zero order in reactants. This is explained as a consequence of the low solubility of the nitrostyrene; the reaction is zero order in ketoester, and because the concentration of the nitrostyrene does not change, has an apparent zero order behavior in that reactant.

Synthesis of ABT-546

With a reliable Michael addition in hand, we initiated the large-scale synthesis of ABT-546.

Nitroketone synthesis

Nitrostyrene **8** was prepared by heating aldehyde and nitromethane together with NH$_4$OAc in HOAc; the product crystallized out of the reaction in 87% yield (eq 3). The ketoester was prepared in a three-step sequence from ethyl dimethylacrylate (eq 4). First, conjugate addition of *n*-PrMgCl was catalyzed by

CuI in the presence of TMSCl at low temperatures. Saponification provided acid, which was transformed to ketoester **7** by first preparing the acyl imidazolide, followed by treatment with potassium ethyl malonate and MgCl$_2$.

The purity of these reaction partners was critical to the success of the Michael addition. It was found that the presence of even 2.5 mol% residual HOAc in the nitrostyrene decreased the rate by over 50%. Likewise, residual carboxylic acid in the ketoester was a stoichiometric poison of the catalyst. Therefore, assays were developed to monitor these levels, and conditions were found which effectively removed the acids.

The ligand was prepared in a straightforward manner. Aminoindanol and diethyl malonimidate dihydrochloride are heated together in THF (eq 5) (9). On cooling, the unsubstituted bis(oxazoline) is precipitated in 81% yield by the addition of 0.5 N NaHCO$_3$. The cyclopropane bridge is installed in 85% yield with 1,2-dibromoethane, employing LiHMDS as the base (10).

Finally, the Michael addition was scaled up (eq 6). When the reaction was run on 13 Mol scale, product nitroketone **9** was obtained in 83% yield, and in 88% ee.

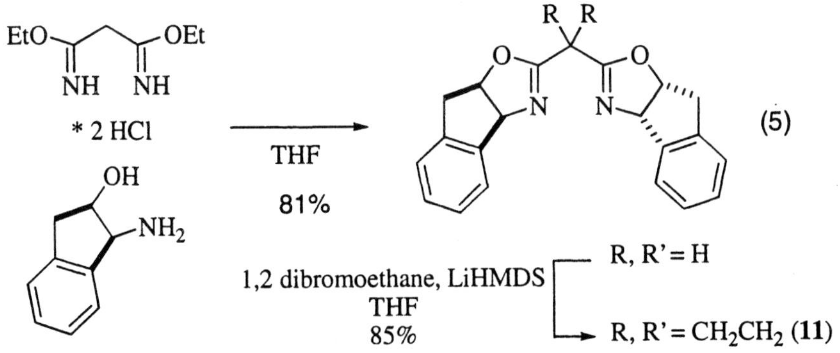

(5)

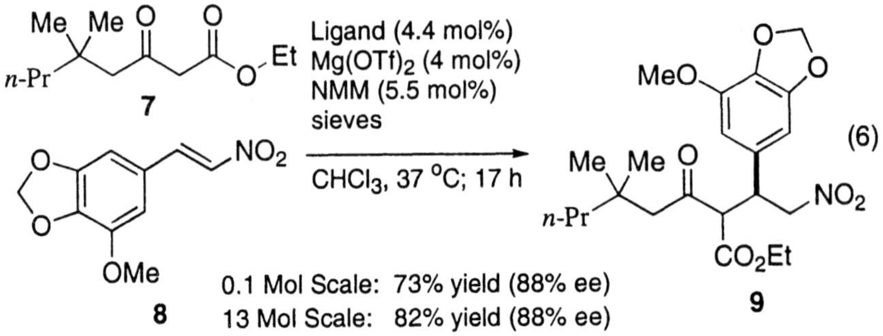

R, R' = H

R, R' = CH$_2$CH$_2$ (11)

(6)

0.1 Mol Scale: 73% yield (88% ee)
13 Mol Scale: 82% yield (88% ee)

Pyrrolidine synthesis

Whereas the nitroketone intermediate in the synthesis of ABT-627 was reduced to the pyrrolidine by treatment first with Raney nickel, then by the addition of TFA to provide the *cis-cis* substituted pyrrolidine selectively, nitroketone **9** gave a mixture of products under those conditions. Modified conditions were identified under which the nitroketone was reduced to the cyclic imine over Raney nickel (Figure 7). Further reduction was carried out with NaHB(OAc)$_3$ to provide the *trans-trans* substituted pyrrolidine directly in 91% yield over two steps, with less than 2% of diastereomeric products.

Finally, the pyrrolidine was crystallized as the tartrate salt. In contrast to the crystallization of racemic pyrrolidine, this crystallization proceeded with good recovery (77%) of >97% ee material. In addition to raising the optical purity, the crystallization provided the first purification from ethyl dimethylacrylate, and the last until the final salt.

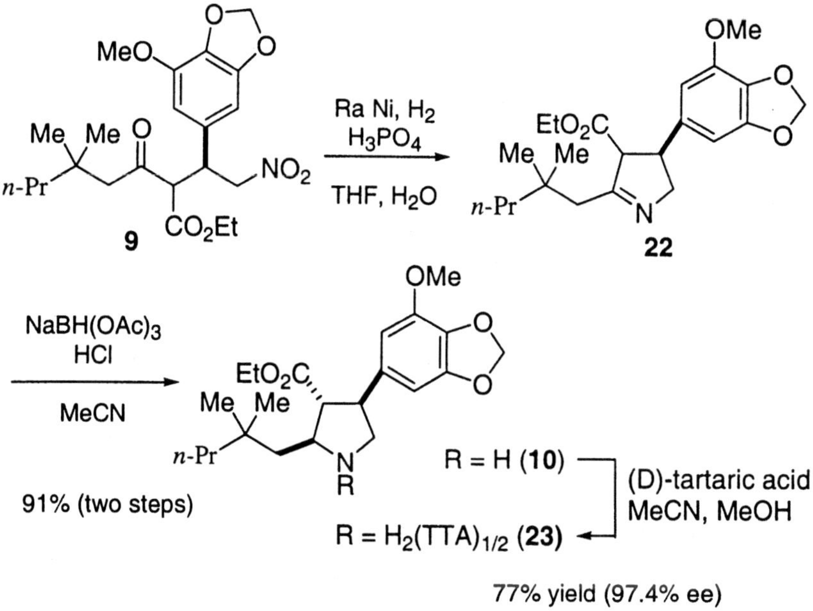

Figure 7. Conversion of the nitroketone to the pyrrolidine-tartrate salt

End game

The final steps of the synthesis were carried out in a one-pot sequence in which the aqueous layer of each step was separated and replaced with the reagents of the following step (Figure 8). The salt was free-based, then the pyrrolidine nitrogen alkylated. Finally, saponification yielded the free-base of ABT-546 in 96% yield for the three steps.

When heptane was added to an IPAC solution of the tosylate salt, the monohydrate of TsOH crystallized out. Therefore, for product isolation, it was imperative that the solution be dried by azeotrope (<6 mol% water). With this precaution, the salt was obtained in 88% yield.

Thus, the synthesis of ABT-546 was accomplished in 11 linear steps from ethyl dimethylacrylate, in 39% overall yield. This represented an 8-fold yield increase from the first generation synthesis.

Scope of the Michael Addition

Following the delivery of 4 Kg of ABT-546 for toxicology and clinical studies, we investigated the scope of the novel Michael addition (Table 1).[11] The reaction of ethyl acetoacetate with nitrostyrene proceeded rapidly at ambient temperature to provide the adduct in 90% ee (entry 1). While the bulkier iso-butyl ester also reacted with high selectivity (88%, entry 2), the *tert*-butyl group proved too bulky and proceeded with significantly lower selectivity

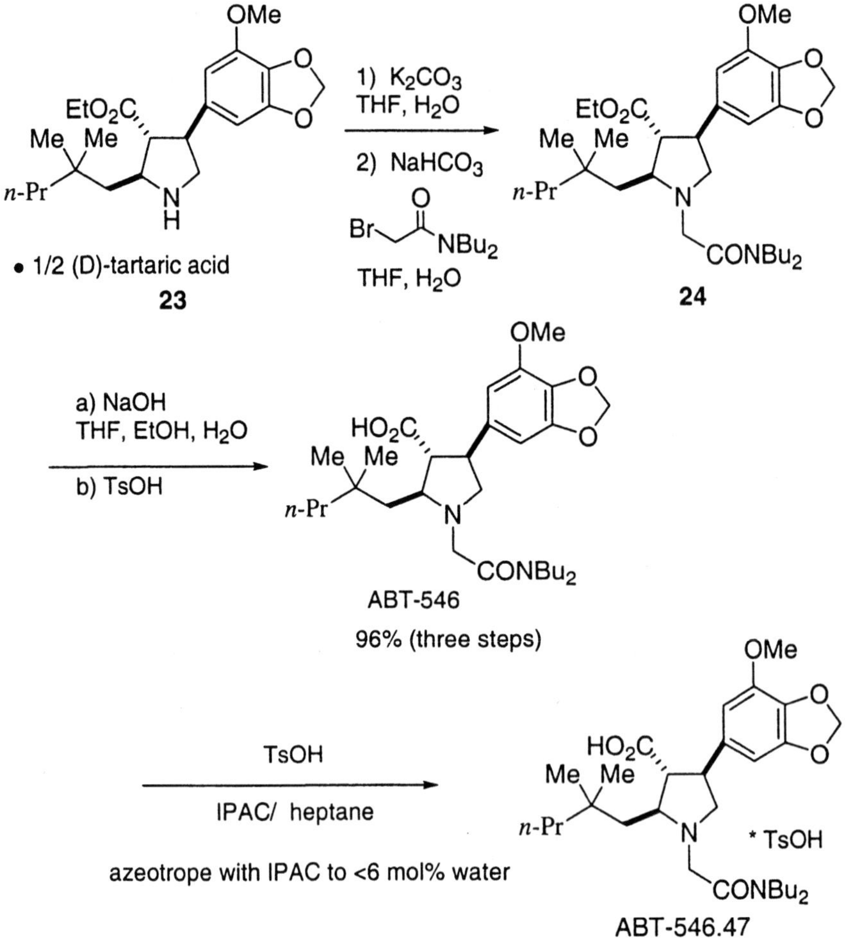

Figure 8. End game

and rate (entry 3). On the other hand, substitution on the ketone moiety is well tolerated (entries 4 and 5).

Table 1. Scope of the asymmetric addition of 1,3-dicarbonyl compounds to nitroalkenes.[b]

entry	R_1	R_2	R_3	selectivity	Yield
1	Me	OEt	Ph	90%	95%
2	Me	Oi-Bu	Ph	88%	92%
3	Me	O$tert$-Bu	Ph	29%	94%
4	CHMe$_2$	OEt	Ph	94%	90%
5	Me, Me n-Pr	OEt	Ph	92%	95%
6	OMe	OMe	Ph	93%	96%
7	OEt	OEt	Ph	95% (93%)[a]	92% (93%)[a]
8	OCHMe$_2$	OCHMe$_2$	Ph	94%	92%
9	OCMe$_3$	OCMe$_3$	Ph	33%	88%
10	OEt	OEt	p-F-Ph	90%	90%
11	OEt	OEt	2,6-(MeO)$_2$-Ph	97% (97%)[a]	93% (95%)[a]
12	OEt	OEt	3,4-(OCH$_2$O)-	93%	98%
13	OEt	OEt	2-furyl	89%	92%
14	OEt	OEt	n-C$_5$H$_{11}$	89%	93%
15	OEt	OEt	Me$_2$CHCH$_2$	90%	88%

[a]Numbers in parentheses refer to reactions in toluene. [b]SOURCE: Reproduced with permission from reference 11. Copyright 1999 American Chemical Society.

The reaction was also extended to malonates, which were generally less sensitive to reaction conditions. As precedented with the ketoesters, the malonate ester group could be varied to a point (entries 6-8), though the *tert*-butyl ester reacted with low selectivity (entry 9). In addition a wide variety of nitroalkenes were effective in the reaction. Electron-poor (entry 10) and -rich (entries 11-12) nitrostyrenes reacted with high selectivity. In addition, the 2-

furyl substituted nitroalkene (entry 13) as well as aliphatic nitroalkenes (entries 14-15) all reacted with equal facility.

Mechanistic studies indicate that the reaction proceeds via initial coordination of the 1,3-dicarbonyl compound to the magnesium center, acidifying the 2-proton. Deprotonation by the amine base results in a chiral magnesium enolate which adds diastereoselectively to the nitroalkene to set the new chiral center.

Summary

In summary, a catalytic asymmetric Michael addition of 1,3-dicarbonyl compounds to nitroalkenes was developed. This reaction was employed in the synthesis of ABT-546 in order to prepare multi-kilogram lots for toxicology and clinical studies. The reaction scope has been found to include a variety of ketoesters, as well as malonates and nitroalkenes.

References

(1) Wu-Wong, J. R.; Dixon, D. B.; Chiou, W. J.; Dayton, B. D.; Novosad, E. I.; Adler, A. L.; Wessale, J. L.; Calzadilla, S. V.; Hernandez, L.; Marsh, K. C.; Liu, G.; Szczepankiewicz, B.; von Geldern, T. W.; Opgenorth, T. J. *Eur. J. Pharm.* **1999**, *366*, 189-201.

(2) Webb, D. J.; Monge, J. C.; Rabelink, T. J.; Yanagisawa, M. *Trends Pharmacol. Sci.* **1998**, *19*, 5-8.

(3) Liu, G.; Henry, K. J., Jr.; Szczepankiewicz, B. G.; Winn, M.; Kozmina, N. S.; Boyd, S. A.; Wasicak, J.; von Geldern, T. W.; Wu-Wong, J. R.; Chiou, W. J.; Dixon, D. B.; Nguyen, B.; Marsh, K. C.; Opgenorth, T. J. *J. Med. Chem.* **1998**, *41*, 3261-3275.

(4) Wittenberger, S. J.; McLaughlin, M. A. *Tetrahedron Lett.* **1999**, *40*, 7175-7178.

(5) For a review see Ferigna, B. I.; de Vries, A. H. M. *Advances in Catalytic Processes*; JAI Press: London, **1995**; pp 151-192.

(6) Sewald, N.; Wendisch, V. *Tetrahedron: Asymmetry* **1998**, *9*, 1341-1344.

(7) The addition of 1,3-dicarbonyl compounds to nitrostyrene has been reported to be promoted by chiral alkaloids in up to 43% ee. Brunner, H.; Kimel, B. Monatsh. Chem. **1996**, *127*, 1063-1072.

(8) In the addition of ketoesters to nitrostyrenes, two stereocenters are formed. The carboxylate-bearing center is formed as a 1:1 mixture of diastereomers, and the selectivity refers to the aryl-bearing center.

(9) The synthesis of the parent ligand in DMF has been reported in 60-65% yield. Ghosh, A. K., Mathivanan, P. Cappiello, J. *Tetrahedron Lett.* **1996**, *37*, 3815.

(10) Davies, I. W.; Gerena, L.; Castonguay, L.; Senanayake, C. H.; Larsen, R. D.; Verhoeven, T. R.; Reider, P. J. *Chem. Commun.* **1996**, 1753.

(11) The material in this section has been communicated. Ji, J.; Barnes, D. M.; Zhang, J.; King, S. A.; Wittenberger, S. J.; Morton, H. E. *J. Am. Chem. Soc.* **1999**, *121*, 10215-10216.

Chapter 5

Process Chemistry: Creativity in Large Scale: Lessons from Scale-Up

Roger N. Brummel

Parke-Davis Pharmaceutical Research, 188 Howard Avenue, Holland, MI 49424

In the pharmaceutical industry, chemical development has been described as carrying out experiments in large tanks. It has also been described as experiments which require a huge amount of money. Yet, in the drug discovery and development process, the majority of the cost incurred in bringing a new chemical entity to the market place do not depend on the cost of producing chemicals on a large scale. Rather, the majority of the cost is funding the clinical trials which take place in late Phase II and Phase III. Consequently, even though the amount of monies and materials that are required in scale-up to produce large batches of active pharmaceutical ingredient (API) is substantial, the cost of the chemical is relatively small compared to the total cost of development.

The Drug Discovery & Development Process Map

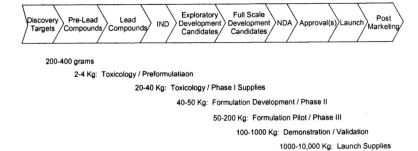

Figure 1: Drug Discovery & Development Process Map

From the discovery and development map, a targeted molecule is originally synthesized by a medicinal chemist and carried on into market by many responsible groups. The development chemist stands midway between the discovery chemist and the manufacturing colleague. Consequently, it is very important that those in development understand the viewpoints that are expressed by research, discovery, and manufacturing. For example, when the medicinal chemist talks about making material, he or she may be talking about making milligrams or hopefully grams of material whereas manufacturing colleagues talk about making tons of material. When ascertaining the amount of raw materials necessary to produce a compound, discovery colleagues are talking about buying materials from a supply house in quantities of grams without regard to cost. However, manufacturing colleagues are talking about establishing independent suppliers that can dependably produce multi-ton quantities of raw materials at a low price. Consequently, it is the responsibility of the development chemist to understand and recognize what is meant when discussing raw materials supply or a finished process producing large amounts of material.

Viewpoints

	Research	Development	Manufacturing
Amounts	5-20 Grams	1-100+Kilos	Tons
Cost	Trivial	Critical	Consistent/Minimal
Purity	Fair	Excellent	Excellent
Purification	Any	Limited	None/Procedures Available
Conditions	Any	Limited	Standard
Raw Materials	Catalog	Bulk	2 Established Suppliers
Waste	Trivial	Critical	Known/Measured
Reproducibility	Moderate	Critical	Exact
Scaleable	Trivial	Critical	Established

Figure 2: Viewpoints

Some of the major considerations to address in defining a commercial process are: (1) How available is starting material? (2) What is the cost of the starting material? (3) What are the hazards or safety concerns? The safety concerns to consider must be from the perspective of the chemical operator who is dealing with the chemicals, the equipment, and the environment.

There is no doubt that as a process is scaled, any undesirable effect is exaggerated. Some effects may have been noted in process development, but others surface as the scale increases. It has been said, "A surprise in research is called a discovery whereas a surprise in development is called a disaster" (*1*). There should be no major surprises in development.

There are two big steps in scale-up. The first is the conversion from the lab to the pilot plant. Ordinarily, a large-scale operation in the laboratory is on the order of 5-22L. Running the same process in the smallest pilot plant equipment would be in 200 to 400 L equipment. This constitutes a 1:30 ratio. On the other hand, moving from the pilot plant to manufacturing involves going from the intermediate to large pilot plant equipment which may be 1000 to 2000 L in size to manufacturing which may be 8,000 L in capacity. This is a ratio of 1:4. Even though this is a smaller ratio, the concerns are large. So, the two big steps in scale-up are from the lab to the pilot plant and then from the pilot plant to manufacturing. Each of those transitions call for new types of observations and new types of solutions to problems.

In the laboratory, reaction times are relatively short. During scale-up, there are longer reaction times and the major concern is the stability of intermediates and final product in the reaction media. Isolation, purification techniques, and times may cause different morphology or crystal structures than noted in the lab. Also, heating and cooling cycles are much longer on large scale than in the laboratory resulting in impurities that were never observed at a small scale.

In the transition from the laboratory to the pilot plant and then to manufacturing, there are many operations which are volume related. In the laboratory, it is very easy to transfer from one flask to another. But in either the pilot plant or in a manufacturing setting, pipes or hoses limit the transfer volume and consequently extends the transfer time. Fluid flow in transferring from one vessel to another is limited by the size of transfer equipment or the capacity of pumps. Also, the transfer of heat or ability to cool depends on the surface area of the equipment. The surface area to volume ratio is less in a pilot plant or manufacturing than in the laboratory.

There are scale-up concerns which are related to operations. In the laboratory, it is easy to observe what is happening. This makes extractions and phase separations a normal operation. We can charge materials and observe frothing or foaming. When the reaction is complete, it is very easy to evaporate solvents on a rotary evaporator. Those same operations are not as easy on large scale in closed vessels where the operator depends on charts, graphs, meters, and instruments.

The most common cause of process variations on a large scale is equipment variations: the material of construction, the shape and size of the equipment, the design of the agitator, the baffle system, and the isolation technique. In controlling or monitoring temperature, the location of the sensors is very important. Is the sensor in the bottom valve or in the baffle? Many things are much different from the laboratory and vary from system to system in the pilot plant or production area. Consequently, as the process progresses from laboratory to pilot plant to production or from one production location to another there are a number of variables which contribute to changes in yield, purity, and physical properties of the desired product.

In the pharmaceutical industry, on the path from discovery to market it is essential that a number of activities occur simultaneously. First, to keep the toxicology and clinical programs going, material must be produced. Second, process research must occur to determine the best commercial route. Third, analytical support and development must occur. The approach, then, is to carry out these activities in parallel as much as possible.

The Parallel Paths Of Chemical Development

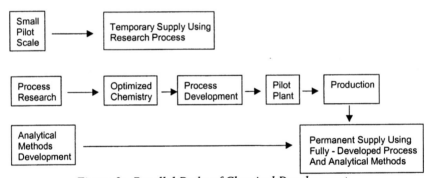

Figure 3: Parallel Paths of Chemical Development

What then is a developed process? For a fully developed process, a number of criteria are met and any attempt to improve it results in a more costly or less safe process. However, in today's environment, the process must not only be

Fully Developed Process

- Process Is Safe
- Chemistry Is Optimized
- Required Analytical Procedures Are Available
- Waste Streams Are Minimized
- Available Equipment Is Used Efficiently
- Cost Objectives Are Met

fully developed, it must be validated. This requires more than optimized yields, proper purity, and meeting cost objectives. In the academic training of a chemist or engineer the goal was to produce the maximum amount of material, the purest material, or the lowest cost material. That became the target. However, in a regulated industrial setting to have a validated process, the critical parameters must be defined. Experiments must be designed to determine the minimums and maximums to produce material which meets specifications.

Process Optimization/Validation

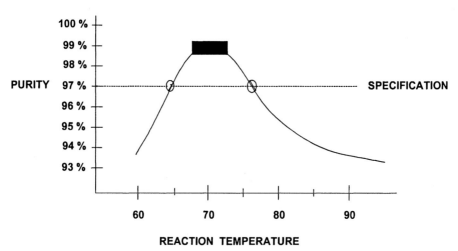

Figure 4: Process Optimization/Validation

The boundary conditions must be identified. For example, in Figure 4, to obtain maximum purity for a given reaction, the process should be run at 68-72°. However, to meet the 97% purity specification, the process can be run within the 63 –76° range. Therefore, a validated process which is fully developed has additional criteria.

A Validated/Developed Process
- Has been scaled
- Has been stressed
- Uses qualified equipment
- Uses readily available starting materials
- Has intermediate and API specifications
- Has been HAZOPED
- Is protected by patents
- Has predictable cost
- Has minimal waste

Within the context discussed and based on the parallel paths of development (Figure 3), first of all, if it is possible, the medicinal chemist's approach would be scaled to produce compound. The rational for this is to supply material to get the drug candidate into humans as soon as possible. The sooner safety and efficacy is determined, the sooner decisions can be made. While producing material using the medicinal chemist's approach, a commercial process can be designed, piloted, demonstrated in manufacturing equipment, and finally validated.

Scaling up the medicinal chemists' approach involves some key issues. It must be done quickly and safely while producing quality product which meets the minimum purity specifications such that programs can continue and early clinical material can be generated. Following are three examples in which the medicinal chemists' approach was used by clearing process hurdles to run at a larger scale and produce material of acceptable quality.

The first example is a reaction which used lithium chemistry (Scheme 1).

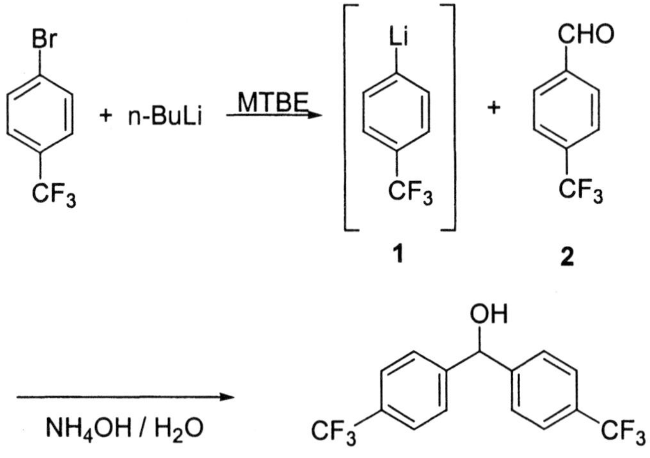

Scheme 1

The medicinal chemist used n-butyllithium on a very small scale and there was no problem producing the lithium salt **1**. This was then reacted with the appropriate benzaldehyde **2** in methyl t-butyl ether to make the required benzhydrol **3**. Based on the kinetics of this reaction and the reactivity of the final product **3**, a continuous reaction reactor system (Figure 5) was made of 3/8 inch pipe fed by dual head pumps such that **1** was reacted immediately with **2**.

This was all done in a -30°C bath and by counter current mixing using baffles in the system. By determining the appropriate residence time, the material was discharged into a quench vessel. Using this system, 1 Kg/hr was produced. Three lots of 15-25 Kg were made. Overall yield was 76% with >99% purity. The project was eventually cancelled. However, the design for the commercial process was to utilize the same reaction system but use larger diameter tubing. As one doubles the radius of the tube, four times the amount of product can be produced.

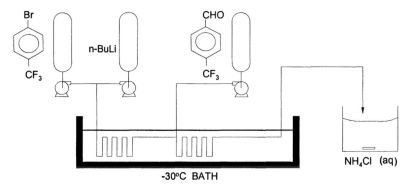

Figure 5: Continuous Lithiation

A second example is a nitration reaction (Scheme 2). This reaction is very exothermic and the nitration product **4** degrades in the reaction medium.

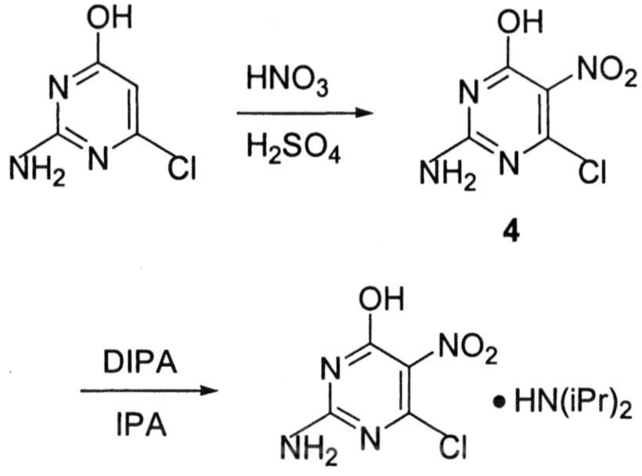

Scheme 2

Because of this degradation, traditional batch reactions always gave poor yield and low purity. A continuous reaction system which controlled the exotherm and immediately removed the product from the reaction media solved the

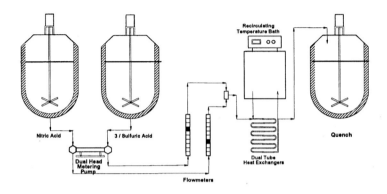

Figure 6: Continuous Nitration

problem (Figure 6). Once the proper reaction residence time was determined, the mixture was quenched into diisopropylamine. This reaction was run on a 25 Kg scale using 90% nitric acid achieving 86% yield. Interestingly, when 98% nitric acid was used, the yield was cut in half even though the material met specifications.

A third system designed to clear the process hurdles that were present in the medicinal chemist's approach was one which used a GC as a thermal chamber. The approach used by the medicinal chemist (Scheme 3) in producing the

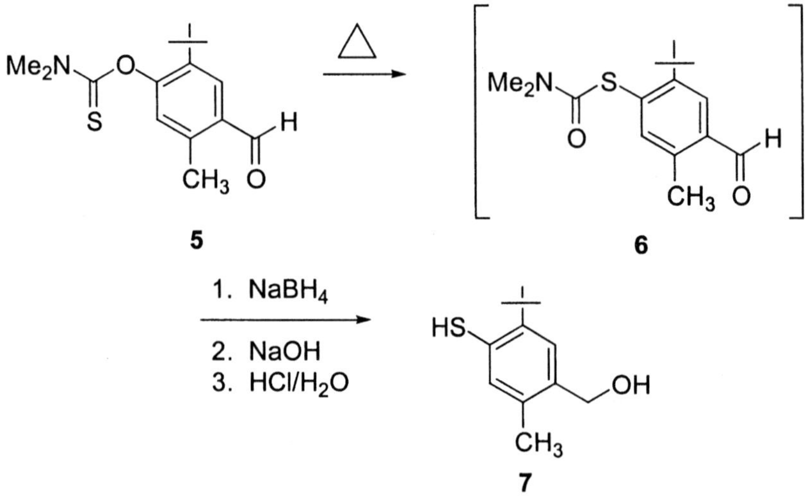

Scheme 3

thiophenol **7** used a Wood's Metal Bath. A Wood's Metal Bath operates at about 270°C but is limited by volume. In the continuous system designed, the thiocarbamate **5**, is injected it into a modified GC at the appropriate temperature, the thermal rearrangement takes place producing the desired thioisocarbamate **6**, is quenched into a sodium borohydride solution, further processed, producing the desired **7** (*2*). (Figure 7) The overall yield was 66% with an HPLC purity > 98%. This

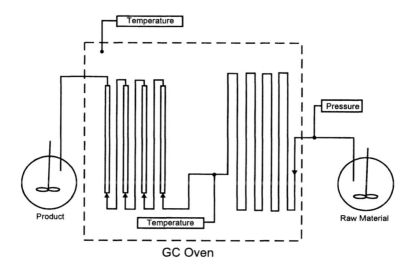

Figure 7: Continuous Pilot Plant Reactor for Thermal Rearrangement

system generated 15 Kg of **7** which allowed the production of material which could be used in late toxicology studies and early clinical trials. At the same time this was being done, another process was developed to intercept one of the targeted intermediates in the overall reaction scheme and the continuous thermal rearrangement was no longer needed. None the less, this method gave the first amount of material which allowed the program to continue.

In process research, a number of potential commercial processes are identified. By piloting candidate processes, a good understanding of processes, variables, and parameters help select the commercial process. Processes for the synthesis of gabapentin **8** had to be piloted for a meaningful choice of a commercial route.

8

Gabapentin

When the volume projections were initially made for gabapentin **8**, it was anticipated that the dose would be low and therefore the volume would be low. Consequently, modifications of the medicinal chemist's approach (Scheme 4) (*3*) were sufficient to produce material and even to consider this as a commercial process.

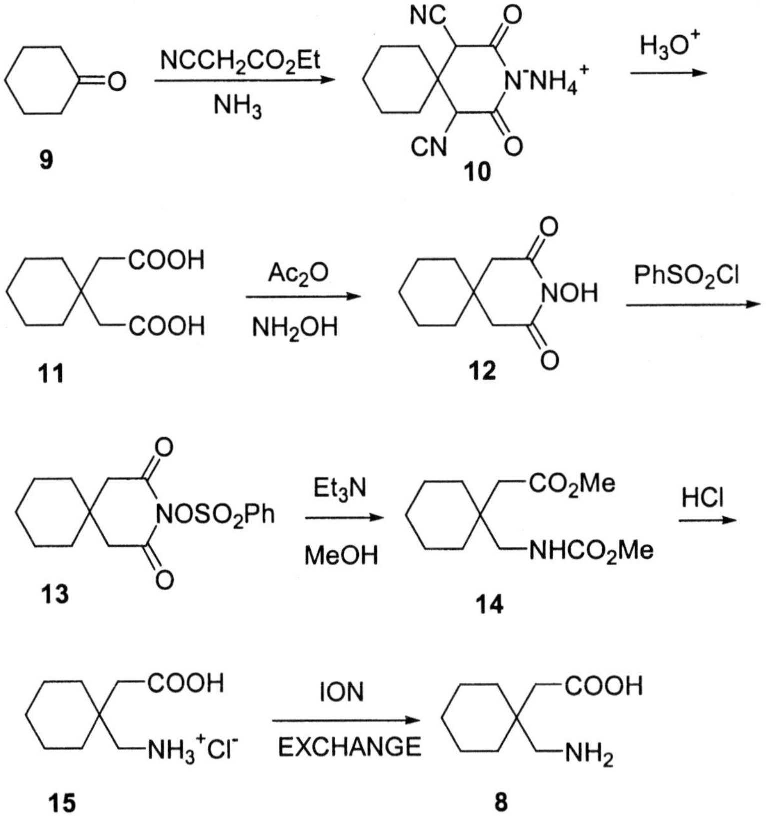

Scheme 4

However, as the clinical studies progressed, the dose escalated and the potential demands increased. Scheme 4 would not be able to be used for commercial production of gabapentin **8**. First was atom efficiency. The simple starting material, cyclohexanone **9**, is built to a molecule **13** with a molecular weight of 337 which is then dissected to the API **8** with molecular weight of 171. This is neither efficient nor esthetically pleasing. Secondly, for the process, the rate-limiting step was converting the hydrochloride salt of gabapentin **15** to the product **8**. The best way to do this was an ion exchange column. However, the maximum concentration of the solution would require substantial equipment to produce the amount of material projected to meet market demands. So, even though this route gave very clean material, it was not atom efficient and had a throughput problem. So other routes had to be considered which would lead to the target molecule.

The first approach investigated was the use of nitromethane (Scheme 5) (4). However, in the conversion of cyclohexanone 9 to final product 8 two things

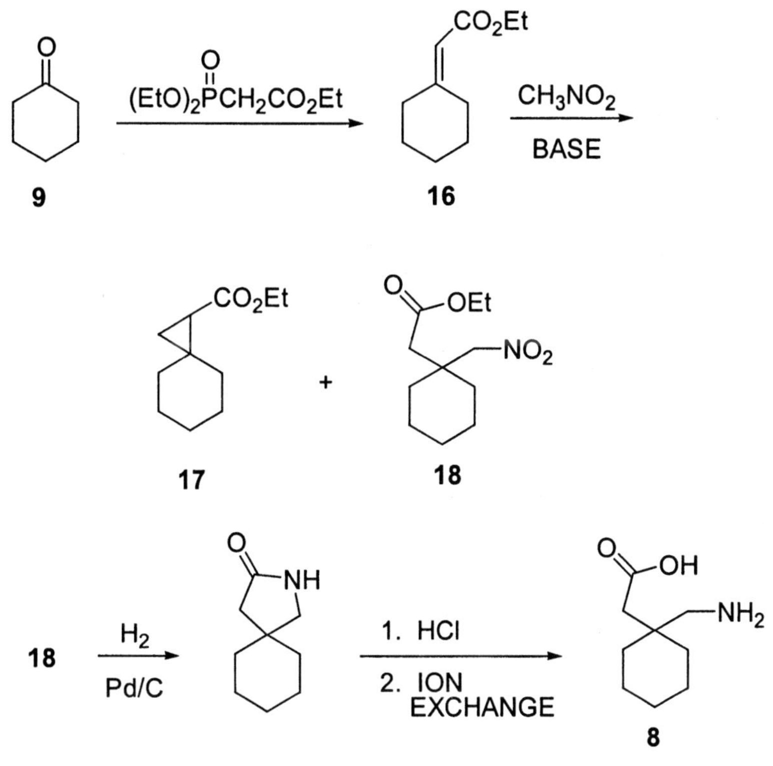

Scheme 5

became apparent: (1) the addition of nitromethane to 16 gave considerable amount of a cyclo propane adduct 17 (5) as well as the desired ester 18 and (2) a hazards evaluation determined the thermal characteristics of this reaction would be a major concern at large scale. Consequently, this route was not pursued even though on small scale, and with modifications it produced excellent material.

Another approach used chemistry described by Scaefer (6) in which a dinitrile **19** was produced by the reaction of cyclohexanone, ethyl cyanoacetate, and cyanide ion. The nitriles can be differentiated and selectively converted to the ester and nitrile **20** by the Pinner reaction (Scheme 6). After ester hydrolysis

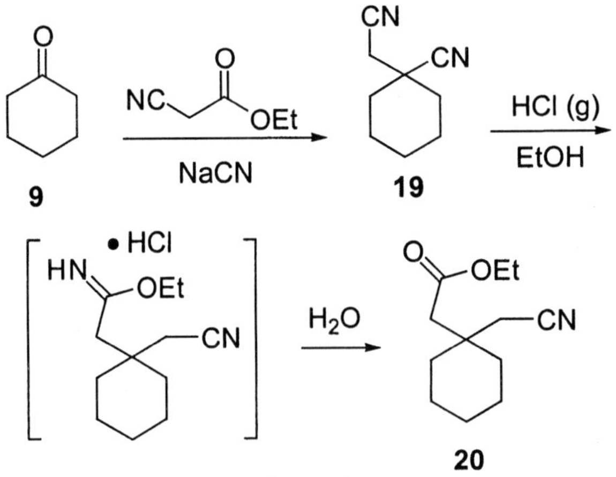

Scheme 6

to the acid **21**, reduction of the remaining nitrile to the amine gave the desired compound **8** (Scheme 7). Initially, rhodium was used as a catalyst but when the price of rhodium became excruciatingly high, it was found that sponge nickel worked better with an increased ratio of gabapentin **8** to lactam **22** in high yield.

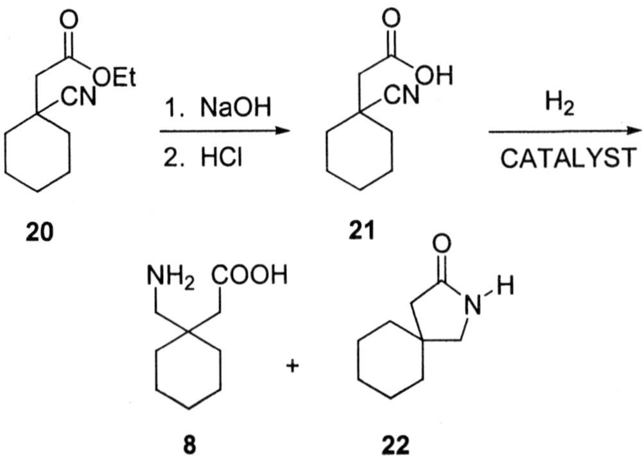

Scheme 7

Consequently, by scaling each one of these reactions, the proper data was obtained to make a choice of commercial route. The Pinner approach (Schemes 6,7) was commercialized and presently is producing more than 100 metric ton per year, in good yield, good purity, and meeting volume demands.

Another example of choosing a route by scale-up in the pilot plant is the case of a hydroxynitrile **25**, a starting material in the atorvastatin process. Described are two approaches that were piloted before a choice was made.

The first uses an inexpensive, readily available simple starting material: starch. In addition to starch, lactose or maltose can also be used (*7,8*). The oxidation of starch by peroxide forms the hydroxylactone **23** preserving the chirality inherent from the starch. Through a series of reactions (Scheme 8), the

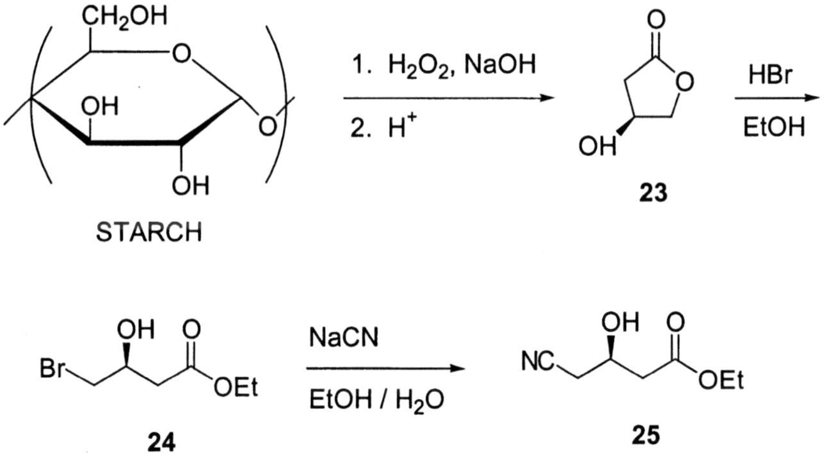

Scheme 8

ring is opened and the cyanide introduced to give the hydroxynitrile **25** in good chemical yield and high optical purity.

One can get into the same reaction manifold by starting with the salt of isoascorbic acid **26** (Scheme 9) (*8,9*). The sodium salt is a common meat preservative and is available at a low price. Reaction of the salt with hydrogen peroxide followed by treatment with hydrogen bromide and acetic acid gives the dibromo intermediate **27** which is then hydrogenated to selectively debrominate to produce **24**. Substitution of the bromide with cyanide gives the hydroxynitrile **25** and again in the desired stereochemical purity. Both of the processes were run a number of times in the pilot plant at 100-200 Kg scale and based on projected economics, throughput, and purity requirements the isoascorbic acid route was chosen.

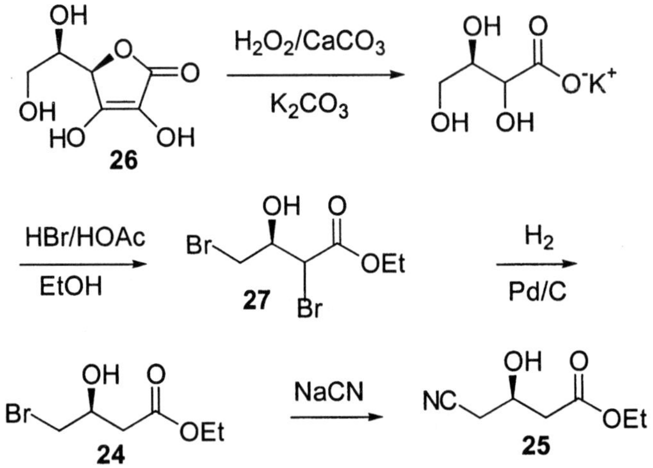

Scheme 9

There are also other ways to synthesize **25** which use biocatalytic systems or chiral hydrogenation catalysts (Scheme 10). Recognizing that suppliers or

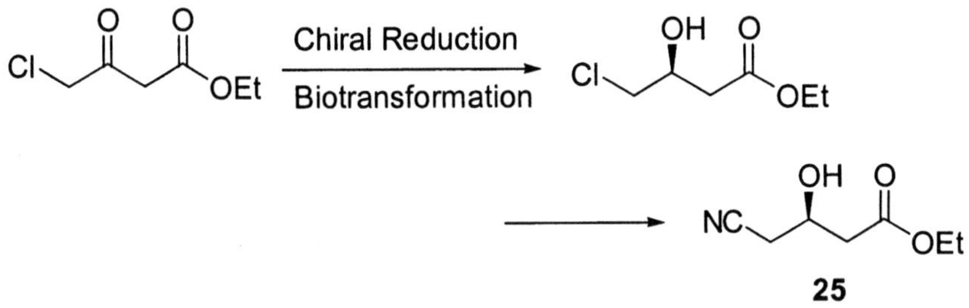

Scheme 10

contract manufacturers may have this or some other expertise, the specifications were set to include these and other approaches to make the hydroxynitrile **25**.

The piloting of commercial processes may not always uncover all the processing problems. For the manufacture of gabapentin **8** and reduction with sponge nickel (Scheme 11), everything worked well in the pilot plant and seemed to work well at one manufacturing location. However, when the process was transferred to another manufacturing location, the reduction would not go. The agitator system at the second manufacturing site was different and not sufficient to disperse the heavy sponge nickel catalyst to achieve the necessary surface contact with the hydrogen in the reduction procedure. It was necessary to add a tickler or extender to the bottom of the agitator to achieve better catalyst/substrate contact.

Another example of things unnoticed in the laboratory or pilot plant was demonstrated in the same reduction (Scheme 11). On large scale in

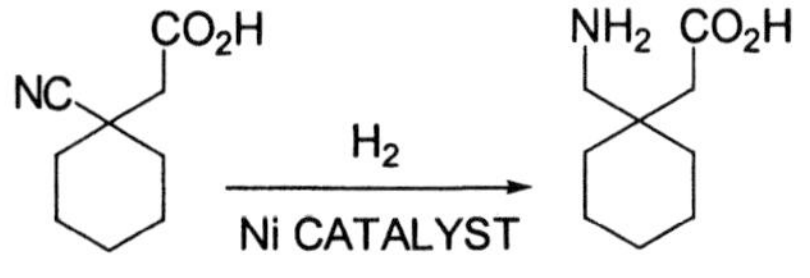

21 → **8**

Scheme 11

manufacturing, after filtering the catalyst, a very blue stream was noted which was never seen in the lab or pilot plant. Analytically, the heavy metal content in product **8** was above specification. In order to reduce the nickel ion content,

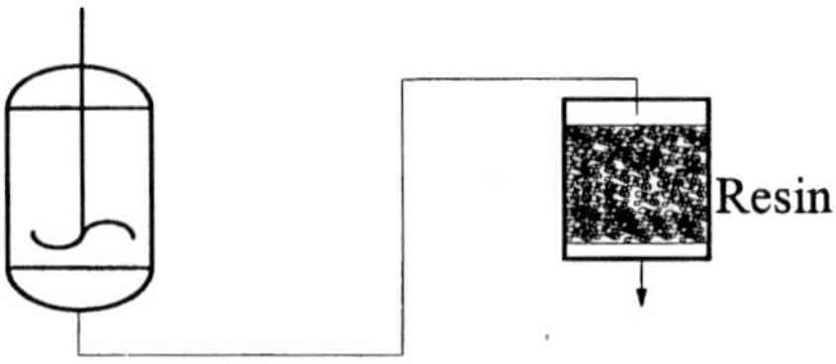

Figure 8: Nichel Removal

it was necessary to pass the stream through a resin (Figure 8) which trapped the nickel and reduced the concentration from over 400 ppm to less than 2 ppm. This was essential in that operation, since the heavy metals specification was 20 ppm or less.

In all processes, consideration must be given to safety. This includes but is not limited to worker exposure. During the development of tacrine **31**, the

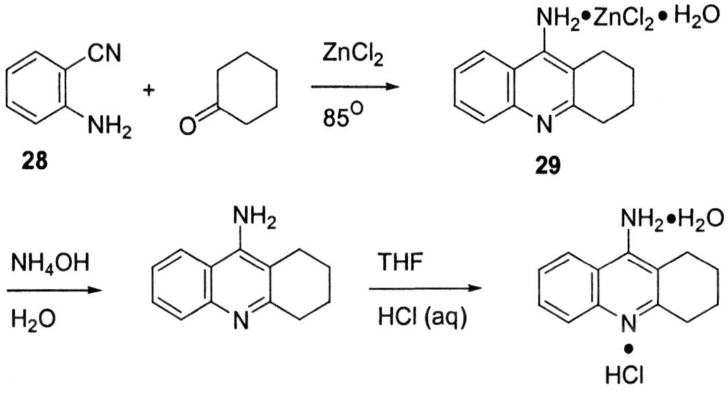

28 + ZnCl₂, 85° → **29** (NH₂•ZnCl₂•H₂O)

NH₄OH / H₂O → **30** (NH₂)

THF / HCl (aq) → **31** (NH₂•H₂O, •HCl)

Scheme 12

industrial toxicology department determined occupational exposure should be minimized. The commercial chemistry developed to be transferred to manufacturing was set (Scheme 12). The overall yield was 90% with purity greater than 98%. However, in the pilot plant scale-up work, the intermediates **29** and **30** had been isolated and handled using extensive personal protective equipment. In production, however, the desire was to use equipment which would contain **29** and **30**. There are a number of commercially available filter dryers which will meet the challenge. A Rosenmund system was chosen. Using that system, the amine benzonitrile **28** was reacted with cyclohexanone in

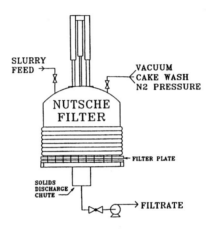

Figure 9: Nutsche Filter

the presence of zinc chloride to produce the intermediate zinc complex **29**. The reaction was done in a normal reactor and the resulting slurry fed into the Rosenmund filter/drier and isolated as a solid **29**. The solid **29** within the filter unit was washed with ammonium hydroxide producing the free base **30** which was also solid. Once **30** was washed of the residual zinc salt, a hydrochloric acid solution is added to form **31**. The API product was also a solid which was dried and discharged into the proper containers and shipped. Using this equipment, exposure was virtually eliminated once the starting materials were charged.

The development of atorvastatin **32** demonstrated the parallel approach used in chemical development. The production of the first kilos of material used modifications of the medicinal chemistry synthesis enabling the process to be carried out in the pilot plant. At the same time, process research designed, developed, and piloted a commercial process which was then transferred to manufacturing.

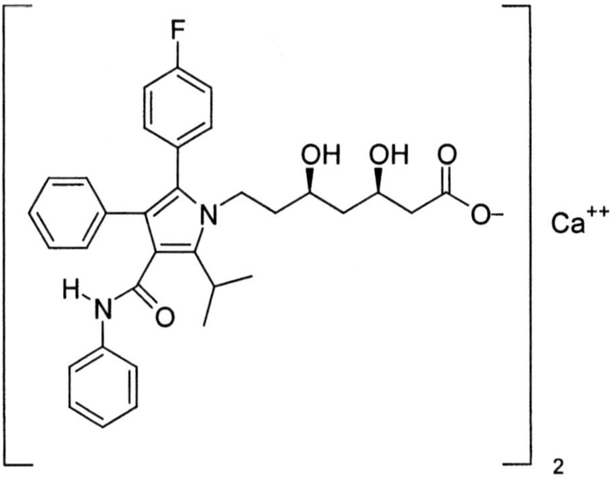

32

Atorvastatin

Retrosynthetic dissection (Scheme 13) of the penultimate lactone defines the parts. The initial method used to produce the first amounts of material

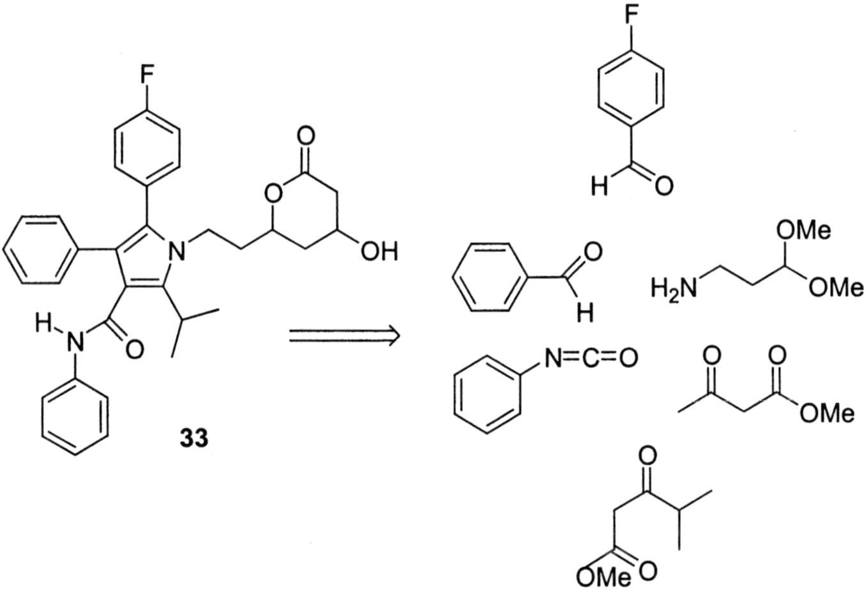

Scheme 13

simply put the pieces together. The synthesis was linear and the choices involved the order of addition *(10)*.

The pyrrole ring **35** was synthesized (Scheme 14) by first making a diketone **34** followed by a Paal-Knorr cyclization. The pyrrole ring **35** then was reacted with phenylisocyanate (Scheme 15) to produce the triaryl pyrrole ring system **36**. Then, through a series of reactions designed to extend the alkyl

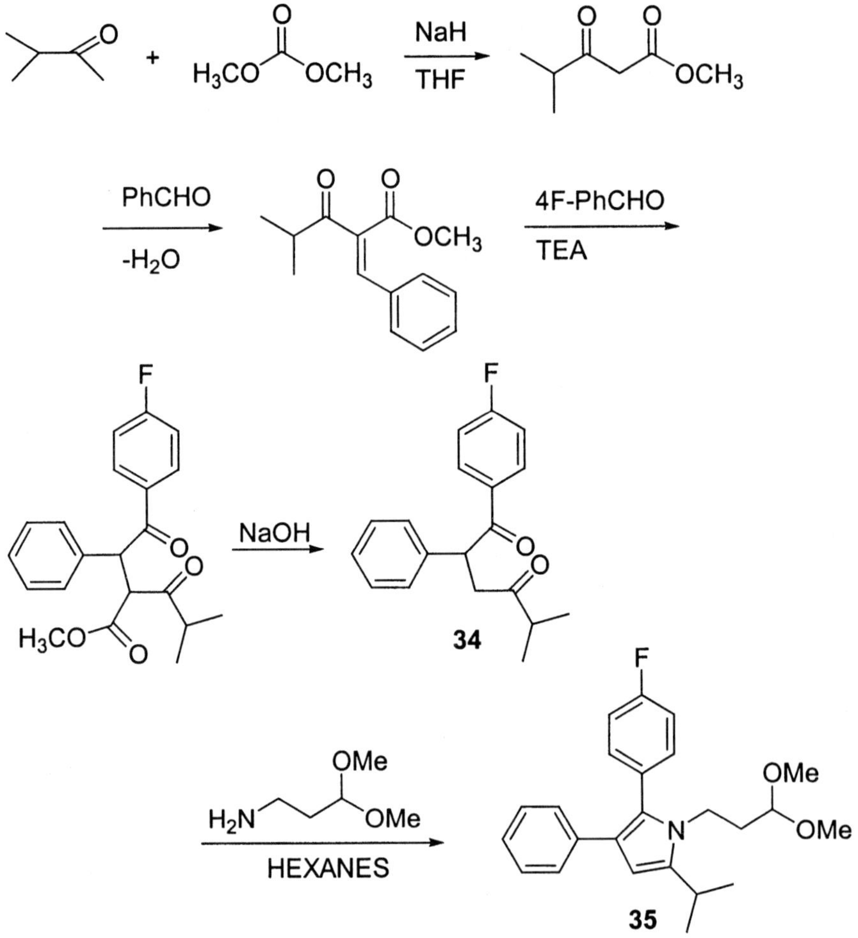

Scheme 14

portion of the molecule, the racemic atorvastatin lactone **33** was produced. This pilot plant chemistry was based on the medicinal chemistry and gave sufficient material to continue the program.

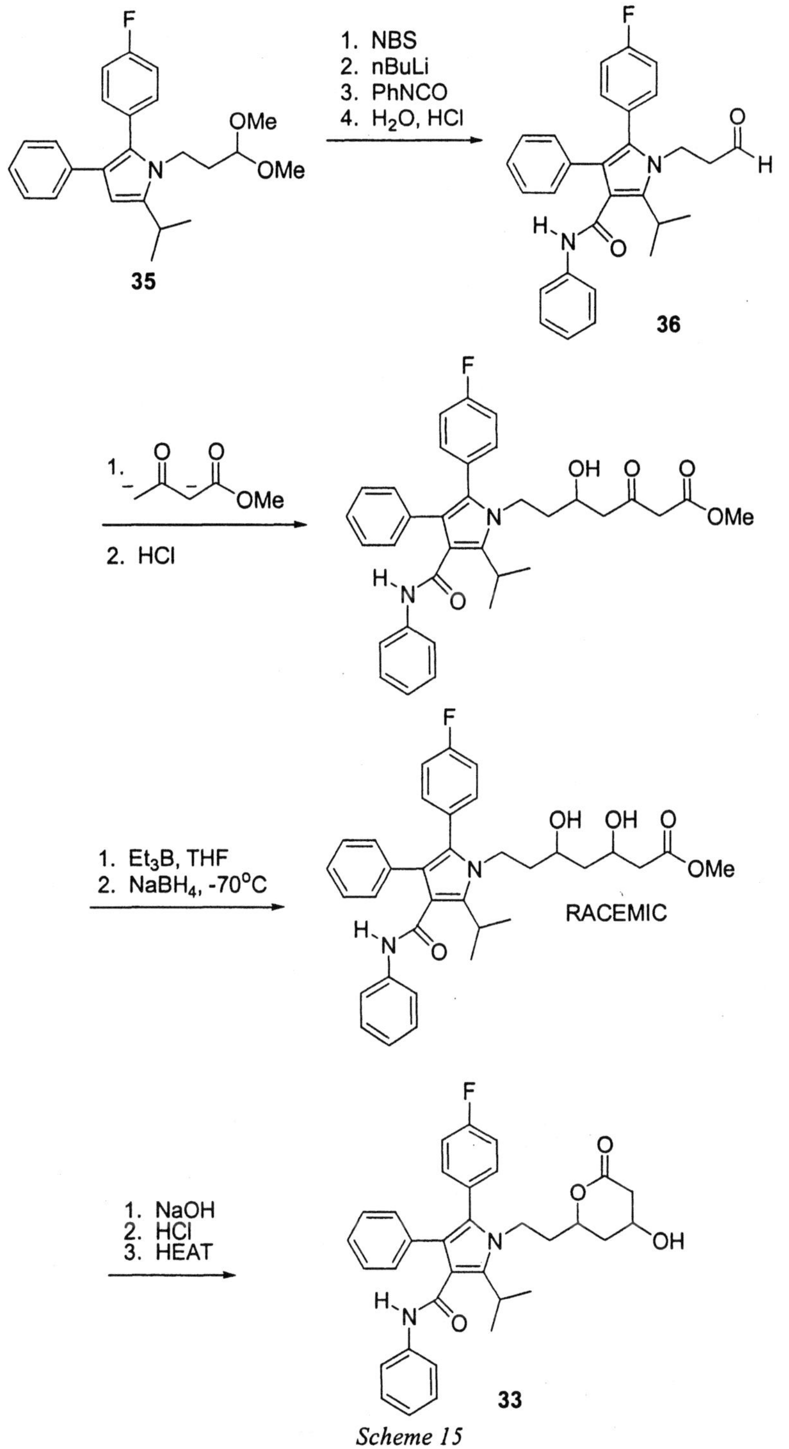

Scheme 15

The design of the commercial process for atorvastatin was a convergent synthesis to produce a chiral molecule. Another look at atorvastatin, **32**, coupled with experiences with the Paal Knorr cyclization, defined the two targets for convergency: the diketone **37** and tert butyl isopropylindene nitrile

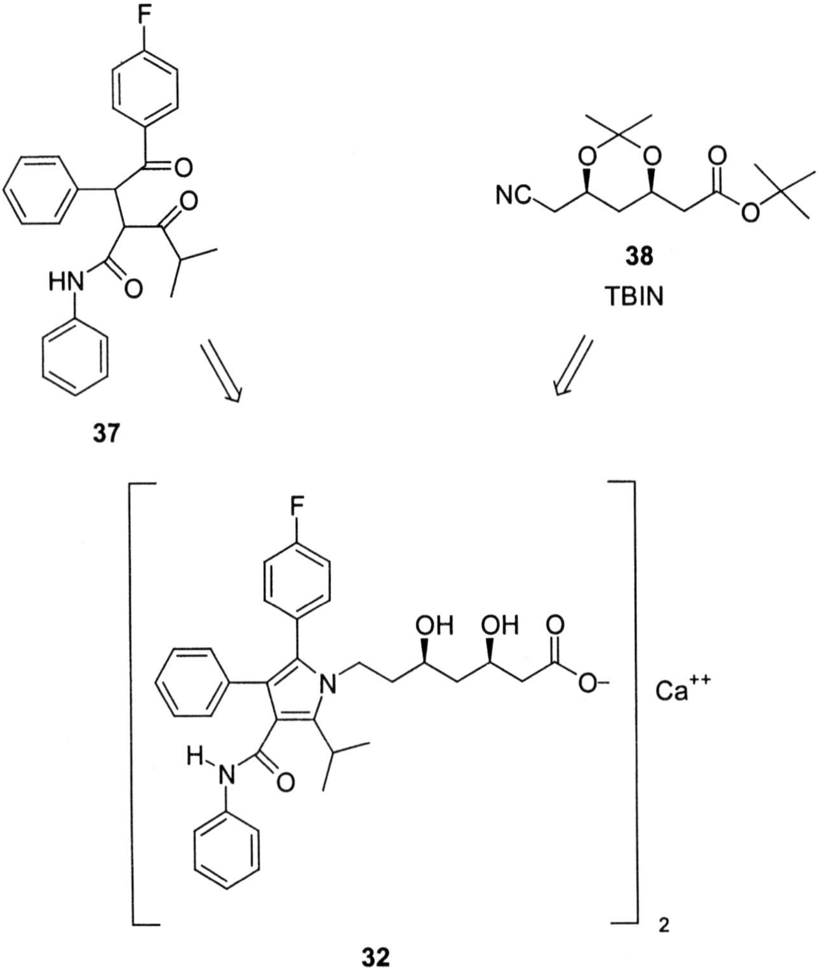

37

38
TBIN

32

Scheme 16

(TBIN) **38**. The diketone synthesis is described in Scheme 17 in which all three aryl rings are added (*11*). The key to the sequence was the Stetter reaction and selection of the proper catalyst (*12*).

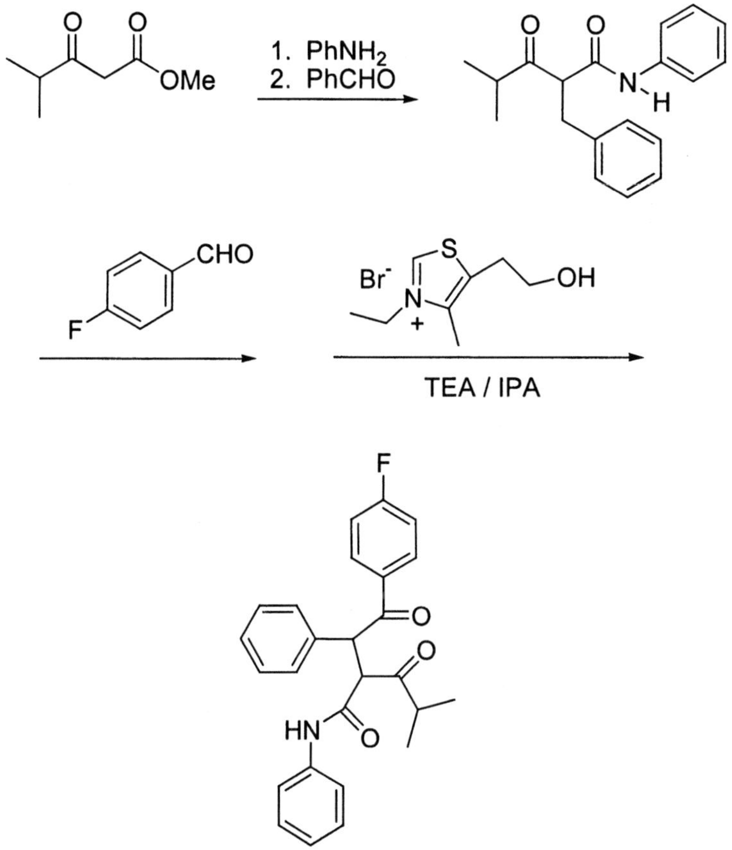

37

Diketone Synthesis
Scheme 17

The TBIN **38** was made (Scheme 18) from hydroxynitrile **25** (*13,14*) which was described earlier. Chain elongation was accomplished through the anion of t-butyl acetate addition to **25**. Selective reduction using triethyl borane and borohydride introduces the second chiral center. Unfortunately, or fortunately, this required low temperature equipment which was not in the stable of normal manufacturing equipment. At least one year was expended trying to run the reaction at warmer temperature. The reaction would go, but the yield was very poor. The formation of lithium t-butyl acetate **39** is run at low temperature, and **25** is added to form the hydroxy ketone **40**. But most important was the correct

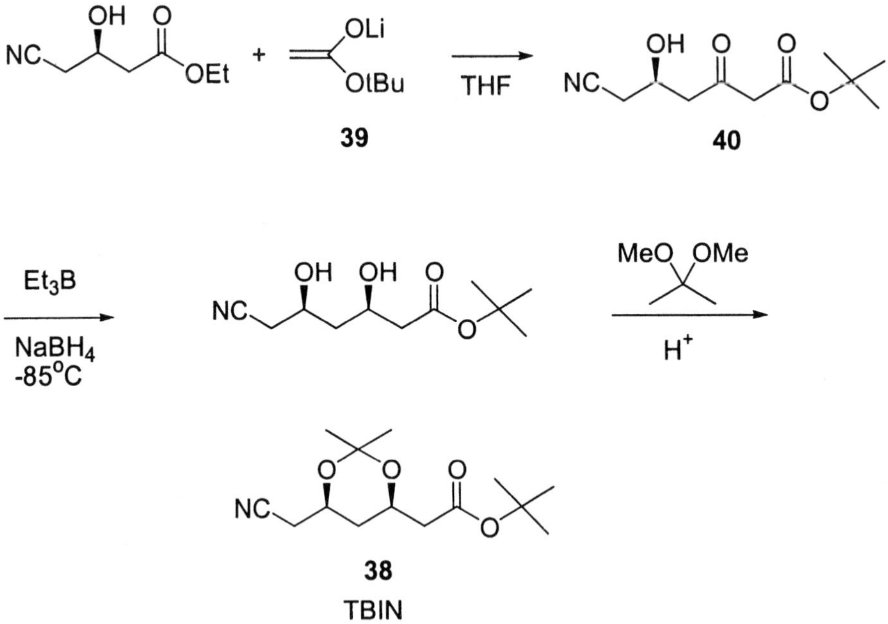

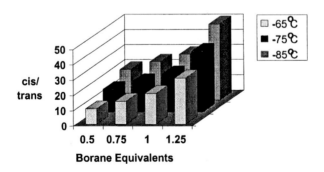

Scheme 18

stereoselective reduction of **40**. The cis-trans ratio as well the yield depends on the temperature of the reaction. Through designed experimentation (Figure 10), it was determined that a 50:1 ratio could be achieved using 1.25 equivalent of borane running at -85°C or lower.

Figure 10: Stereoselective Reduction

This process presently is being run at a 1500 gallon scale and achieving the yield as projected and the chiral ratio that is described. Once the desired chiral dihydroxy compound is produced, it is protected as an isopropylidine **38**.

The TBIN **38** is reduced to the corresponding amine **41** and under Paal Knorr conditions reacted with the diketone **37** (Scheme 19) (*12*). Through experimental design, the yield of the Paal-Knorr cyclization was optimized and presently achieves greater than 70% yield of **42**. Key conditions were defined to

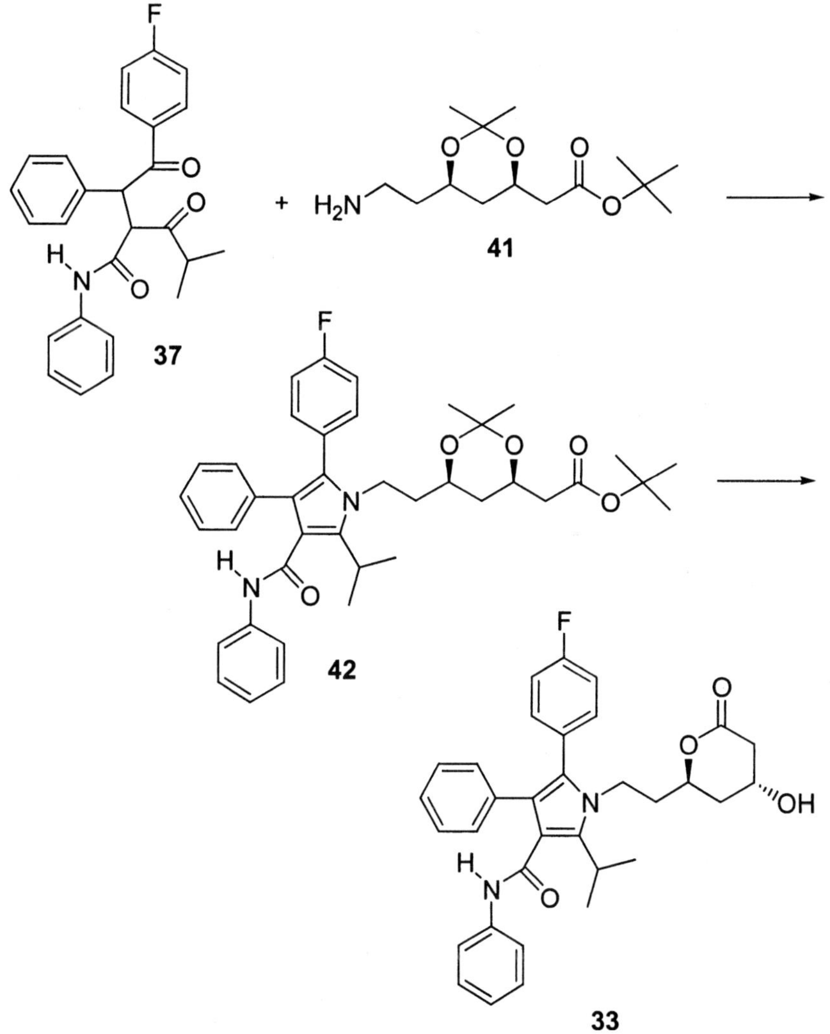

Scheme 19

deprotect, saponify, and form the lactone **33** which is isolated and purified. This has been run on a commercial scale for more than two years obtaining excellent results. The lactone **33** is then converted to atorvastatin **32** in high yield and high purity. The commercial process has produced multi-ton quantities of excellent material and is very robust, having been run with no process failures.

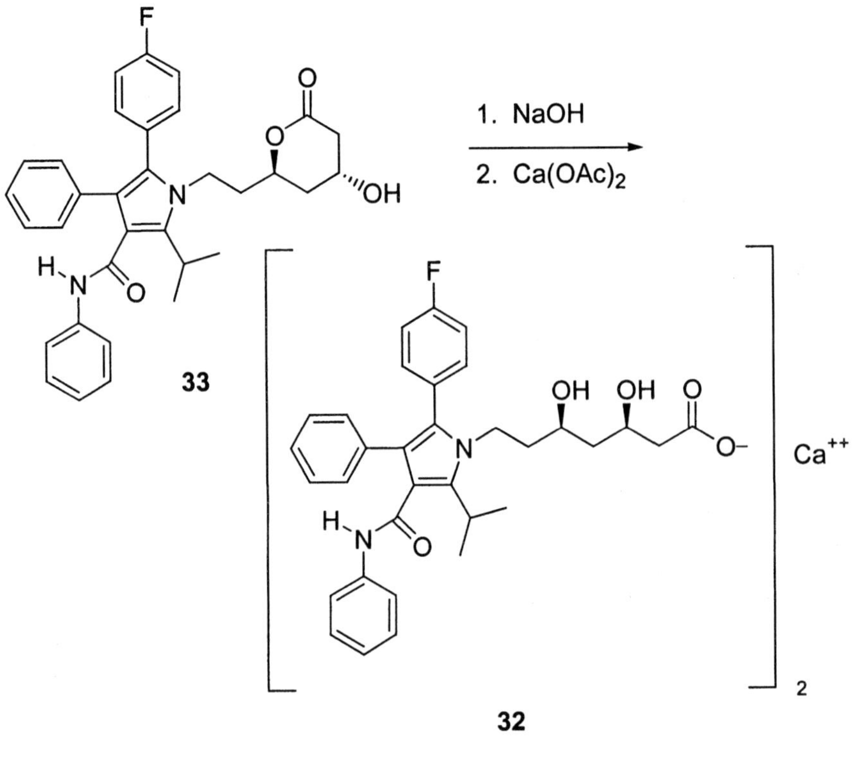

Scheme 20

The previous examples tried to illustrate some events that occur on the parallel paths of chemical development. For each reaction and project, extensive data can be generated – some essential, some nice to have, and some academic. During the development and scale-up of processes, common sense based on sound physical chemistry and engineering principles help to make initial decisions in the plans to transfer from the laboratory to the pilot plant to manufacturing (Figure 11).

Top Ten General Principles of Large Scale Reactions

1. Reaction rates double every 10°C.
2. Vapor pressures double every 20°C.
3. A refluxing system is easier to scale-up than a jacket cooled system.
4. Liquids are usually added to solids.
5. Exotherms are controlled by addition of a reactant.
6. Reactions are easier to inert on large scale.
7. Heat transfers 33% faster in a stainless reactor than a glass reactor.
8. Solid particles tend to collect at liquid-liquid interfaces.
9. Recycle streams usually cause problems after 5 recycles.
10. Solutions cannot be stripped to dryness.

Figure 11: Top Ten General Principles of Large Scale Reactions

It also must be recognized, especially by decision-makers, that to a development scientist, a process is never fully developed. There is always another experiment to try. One should recognize, then, that there are process freeze points where someone must take time to write up the process as if a snapshot was taken.

Process Freeze Points

- Consumer decides that he has no further need for the product
- Project is re-prioritized
- Project is transferred to Manufacturing
- Filing dates: IND, NDA

Obviously, the most desired outcome is to transfer the most elegant, practical, efficient process to manufacturing for regular production. "...an elegant process to an industrial chemist is something to be carried out in a disused bathtub by a one-armed man who cannot read, the product being collected continuously through the drain-hole in 100 percent purity and yield." *(15)*

Acknowledgments

My grateful thanks to Jenny Sloothaak and Sandy Dokter for all the work they did in preparing this paper and the presentation on which it was based.

The examples cited and the work done are a tribute to the creativity, dedication, attitude, professionalism, and spirit that exists in our department. Everyone has contributed to every project, if not directly than indirectly, with ideas, sharing or sacrifice of equipment, sharing of knowledge and experience, and words of encouragement. A special thanks to those listed below who were the persons directly involved and responsible for the specific examples used.

Kelly Baumann
Dan Belmont
Chris Briggs
Don Butler
Jim Davidson
Carl Deering
Randy De Jong
Gary Dozeman
Victor Fedij
Phil Fiore
Randy Geurink
Wolfgang Herrmann
Marvin Hoekstra
Wilfried Hoffmann
Tom Jacks
Rex Jennings

Tung Le
Edward Lenoir
Rick McCabe
Ken Mennen
Alan Millar
Brian Moon
Lyle Mulder
Tom Nanninga
Tim Puls
Bruce Roth
Klaus Steiner
Brian Swierenga
Robert Wade
Jim Wemple
Joachim Witzke
Jim Zeller

References

1. Gardner, C.
2. Lin, S.; Moon, B.; Porter, K.; Rossman, C. A.; Zennie, T.; Wemple, J.; *A Continuous Procedure for Preparation of Para Functionalized Aromatic Thiols Using Newman-Kwart Chemistry; In Situ Generation of Tosyl Bromide for Use in Preparation S-Arylthiosulfonates; submitted for publication.*
3. Hartenstein, J.; Satzinger, G.; U.S. Patent 4,152,326, 1979.
4. Witzke, J.; Hoffmann, W.; EP 414275-A, 1992.
5. Tamura, R.; Kamimura, A.; Ono, N.; *Synthesis* **1991**, p 423.
6. Scaefer, H. *Liebigs Ann Chem* **1965** Vol. 688, p 113-21.
7. Richards, G. N.; Machell, G. *J. Chem. Soc.* **1960**, p. 1924.
8. Hollingsworth, R. I. *J. Org. Chem.* **1999** Vol. 64, p. 7633.
9. Isbell, A.; Frush, H. *Carbohydrate Research* **1972** Vol. 72, p. 301-304.
10. Butler, D. E.; Deering, C. F.; Millar, A.; Nanninga, T. N.; Roth B. D.; U.S. Patent 5,124,482, 1992.
11. Baumann, K. L.; Butler, D. E.; Deering, C. F.; Mennen, K. E.; Millar, A.; Nanninga, T. N.; Palmer, C. W.; Roth, B. D.; *Tetrahedron Letters*, **1992** Vol. 33, p. 2283.

12. Brower, P. L.; Butler, D. E.; Deering, C. F.; Le, T. V.; Millar, A.; Nanninga, T. N.; Roth, B. D. *Tetrahedron Letters* **1992**, Vol. 33, p. 2279.
13. Brower, P. L.; Butler, D. E.; Deering, C. F.; Le, T. V.; Millar, A.; Nanninga, T. N.; Roth, B. D.; U.S. Patent 5,155,251, 1992.
14. Millar, A.; Butler, D. E.; U. S. Patent 5,103,024, 1992.
15. J.W. Cornforth, Nachr. Chem. Tech. 24, 34, 1976

Chapter 6

Synthesis of RO 113-0830, a Matrix Metalloproteinase Inhibitor: From Research Scheme to Pilot-Plant Production

Lawrence E. Fisher[1], Charles Dvorak[1], Keena Green[1],
Samantha Janisse[1], Anthony Prince[1], Keshab Sarma[1],
Paul McGrane[1,3], David Moore[1,3], Jeffrey Campbell[2], Janel Baptista[2],
Chris Broka[2], Than Hendricks[2], Keith Walker[2], and Calvin Yee[2]

[1]Process Research and Scale Up, Roche Bioscience,
3401 Hillview Avenue, Palo Alto, CA 94304
[2]Inflammatory Diseases Unit, Department of Medicinal Chemistry,
Roche Bioscience, 3401 Hillview Avenue, Palo Alto, CA 94304
[3]Current address: CellGate, Inc., 552 Del Rey Avenue,
Sunnyvale, CA 94086

RO 1130830 (**1**, figure 1 below), currently under investigation for the treatment of osteoarthritis, is prepared on a multi-hundred-kilo scale in four steps from advanced intermediates supplied from commercial sources. Crucial to the success and timeliness of the scale-up was close collaboration between research, chemical development and scale-up, pilot plant production and commercial sourcing. Identification of key bond formation sites, reduction of the longest linear sequences and minimization of the number of isolations advanced evolution of the synthetic scheme from the initial medicinal chemistry route to a scaleable pilot plant process.

Process Chemistry

The mission of the process group at Roche Bioscience is twofold: supply active pharmaceutical ingredient (API) of the appropriate quality in necessary quantity to meet the aggressive time lines established by the clinical units within Roche; and, as early as possible, to discover and develop the chemical róute and early process for compound supply which will be used to manufacture the drug candidate through launch. In order to meet these lofty goals, we necessarily must work in the seemingly incompatible areas of process research and compound supply, while complying the ever tightening rules of cGMP which regulate the production of API. The compound and project about which this chapter is written illustrates how the two parallel activities of supply and process research can successfully satisfy the requirements of the clinical groups and "downstream" production/supply.

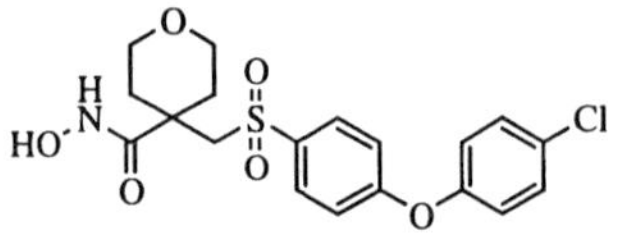

Ro 113-0830
1
Figure 1: Structure of Ro 113-0830

Medicinal Chemistry Synthesis

When presented with the medicinal route (figure 2) several issues immediately stood out: number of linear steps, chromatography, low temperature reactions, low to moderate yields, distillation for purification and use of expensive reagents. Since the presentation of a compound to our group carried with it the expectation of near immediate supply of API for advanced *in vitro* and *in vivo* testing, foremost in our thoughts was to develop a route, to set the final synthetic steps and produce 100+g quantities quickly.

Final API Production Step

The final step of the synthesis is the most critical one to set early because it has the largest, most direct effect on API quality. Oxidation of aryl sulfide **10** to sulfone **1** can be accomplished in a number of ways.[1] At the outset, oxone as an oxidant had several disadvantages (high molecular weight, low water solubility, and large amounts of $KHSO_4$ waste) which tempered its potential

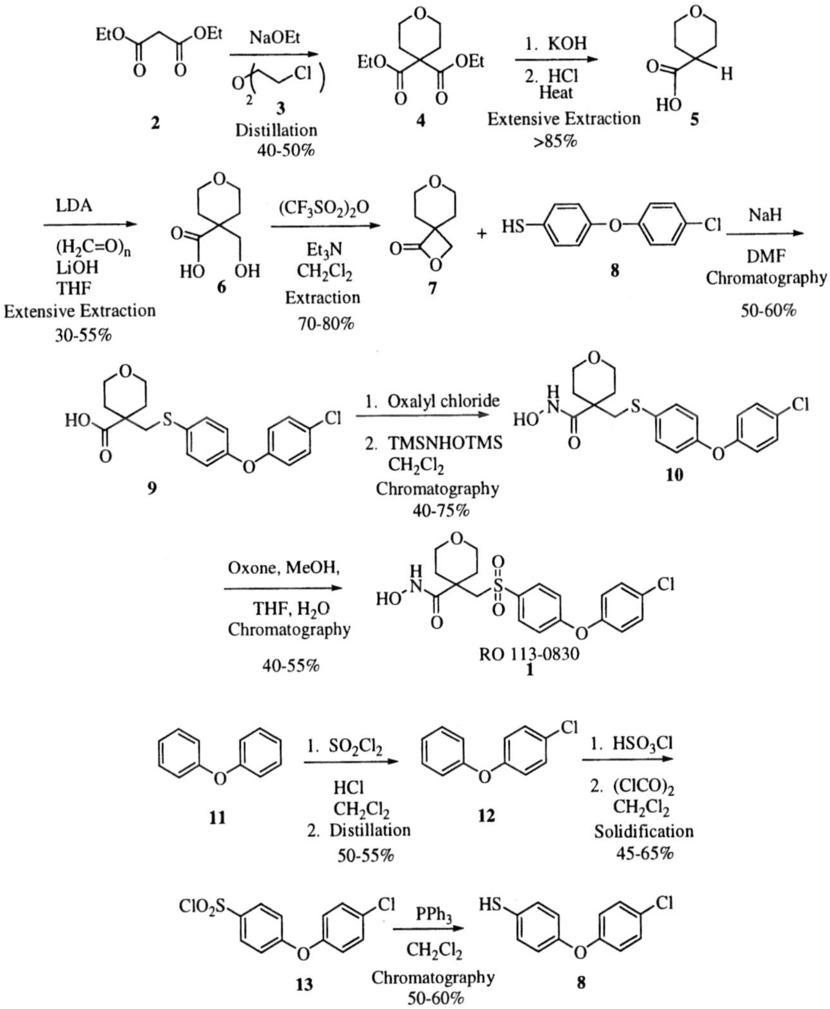

Figure 2: Medicinal route of chemical synthesis

usefulness to the long-term synthesis of **1**. Although reversing the order of the steps by oxidation of carboxylic acid **9** and installation of the hydroxamic acid moiety was examined; it was found to be inferior to the initial sequence. Therefore, due to the immediate need for API, a workable procedure using oxone was developed.

The final API was found to have low solubility in THF/MeOH. This fact, combined with a purity requirement of greater than 98% of the final product dictated changing the penultimate solvent from THF/MeOH to NMP/H_2O. Switching to this solvent system raised the yield, increased throughput and obviated chromatography. This oxidation was observed to have an induction period, and two distinct time components: oxidation to the intermediate sulfoxide, a fast reaction, and final slow oxidation to the product sulfone. Safety concerns led to the decision to perform this reaction at an elevated temperature to eliminate the possibility of a runaway reaction. This also fortuitously allowed less solvent to be used to keep the relatively insoluble sulfoxide in solution. Ultimately, this oxone step was used to produce enough API to satisfy the early and intermediate clinical program.

As time allowed, other oxidation conditions (figure 3) were tried. Performic acid formed in situ using formic acid and hydrogen peroxide ultimately replaced oxone as the oxidation condition of choice in large equipment.[2]

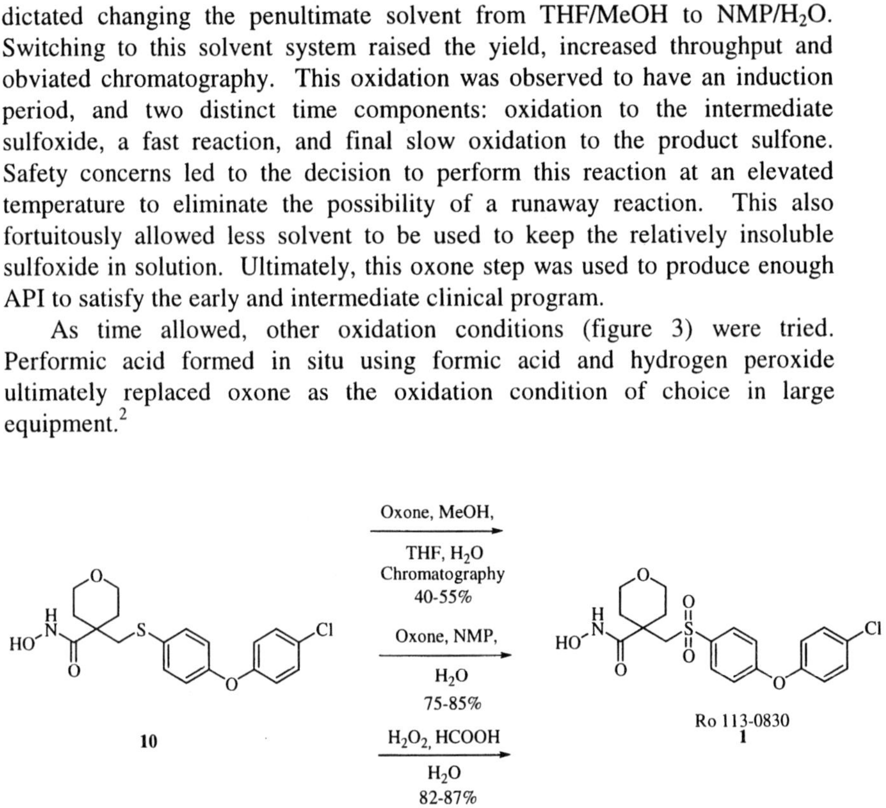

Figure 3: Final step

Final Intermediate Issues

Critical to ensuring high quality API is establishing the quality of the final isolated intermediate. Not surprisingly, high quality (>97% Area Normalization (AN) HPLC) hydroxamate was required to ensure appropriate quality of API. This inevitably required finding acceptable crystallization conditions for 10. The hydroxamic acid moiety in 10 was originally prepared from bistrimethylsilyl hydroxylamine and the acid chloride of 9 (prepared by the reaction of 9 with oxalyl chloride, or less efficiently with thionyl chloride). While the quality of hydroxamic acid was uniformly high using this approach, it was clear that a less expensive, simpler route needed to be developed, due to the intractable, expensive and capricious nature of bistrimethylsilyl hydroxylamine.[3,4] In addition, a simple, robust crystallization method was essential to assure high quality, "usable" 10.

After a number of highly unsatisfactory attempts to develop a procedure to prepare bistrimethylsilyl hydroxylamine from chlorotrimethylsilane and 50% aqueous hydroxylamine, the obvious was attempted: reaction of aqueous hydroxylamine with the acid chloride of 9, generated in situ in acetonitrile. This gave satisfactory 10, which could be crystallized at 80-90 °C by the addition of water (figure 4). Material from these conditions >98% pure by AN HPLC) for direct conversion to API.

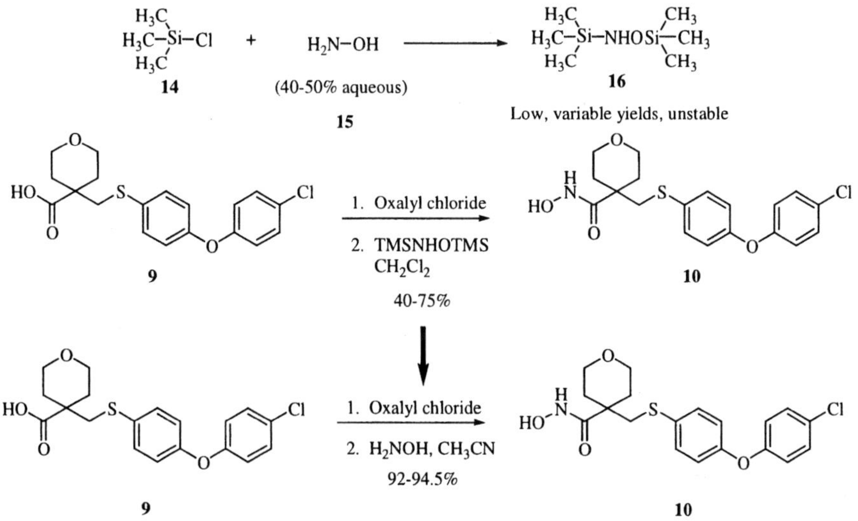

Figure 4: Hydroxamic acid preparation

Critical Bond Forming Reaction

The choice of bond formation to synthesize **9** was addressed in parallel with the activities just discussed. In point of fact, parallel activities were proceeding to loosely follow the medicinal chemistry route to **9**, eliminating chromatography and isolations as possible. The retrosynthetic disconnections shown (figure 5) were examined in order to develop strategies for a concise, short synthesis of **9**.

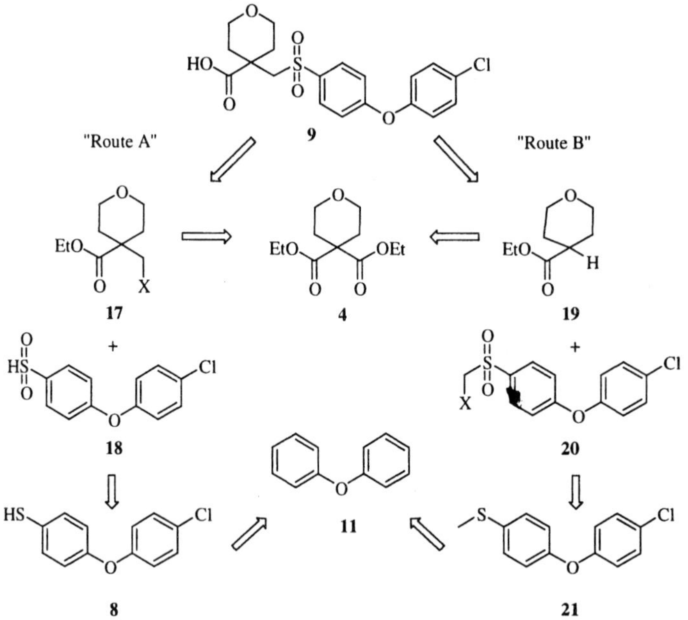

Figure 5: Retrosynthetic disconnections

Retrosynthetic disconnection of **9** "Route A" represents the medicinal chemistry route. It was generally followed through the early stages of this project because project timelines were so tight. However, several significant changes were made to the critical electrophilic intermediate, propiolactone **7**. This material and the intermediates leading to it presented material handling problems as well as isolation challenges. Therefore, tosylate **25** was targeted. A short synthesis of **25** was developed (figure 6). Simple tosylation of alcohol **22** from the DIBAL reduction of diester **5** gave **25**.[5] Interestingly, **28**, a synthetic equivalent of **25** could be produced directly from monoester **26** obtained simply by decarboxylation of **4**. Treatment of **26** successively with LDA and bromoiodomethane gave **28** in one "pot" in 65-70% yield.

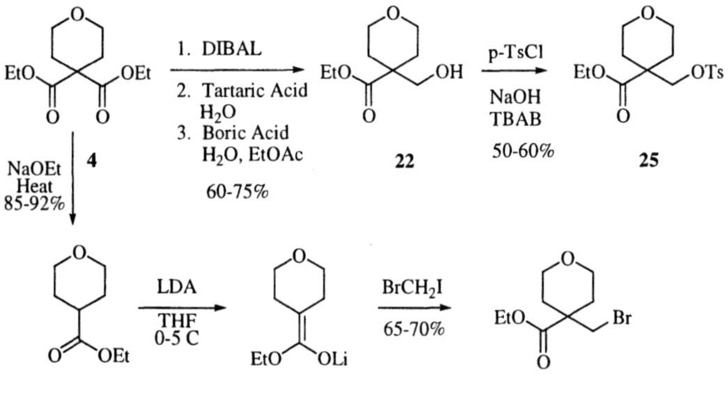

Figure 6: Synthesis of ***25***

Thiol **8** was reacted with tosylate **25** in any of several solvents led to acid **9** (after base mediated ester hydrolysis) in 75-85% yield. Bromomethyl ester **28** also reacted smoothly with **8**, producing acid **9** in 70-75% yield after *in situ* base mediated hydrolysis (figure 7).

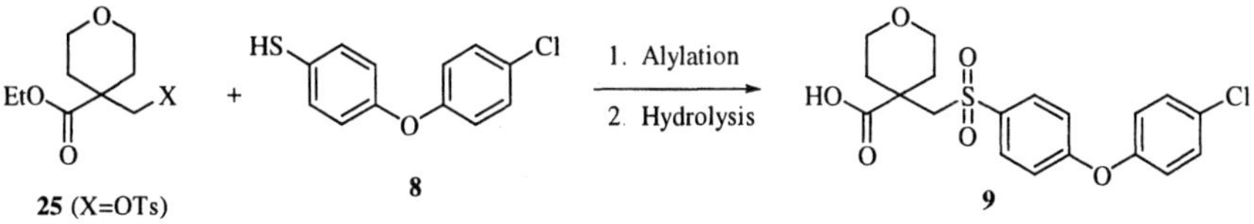

25 (X=OTs)
28 (X=Br)

8

9

75-85% from **25**
70-75% from **28**

Figure 7: Pyran alkylation and ester hydrolysis

Monoester **26** presented an opportunity to explore the bond disconnection in route "B" (figure 8). Anion **27**, generated as before from the action of LDA on monoester **26** reacted readily with chlorotrimethylsilane to give silylketene acetal **29**. This ketene acetal reacted smoothly with chloromethylsulfide **30** in the presence of $ZnBr_2$ in THF at ambient temperature to give ester **9a**, which was hydrolyzed *in situ* to acid **10**.[6]

Silicon waste presents a serious problem for disposal. It tends to coat all exhaust surfaces and not incinerate well. Fortuitously, the lithioanion **27** reacted quite smoothly to give **9a**, obviating the use of TMSCl and getting rid of the onerous silicon waste stream. *In situ* hydrolysis of **9a** and crystallization of **9** from ethyl acetate provided material of sufficient quality to proceed directly to **10**.

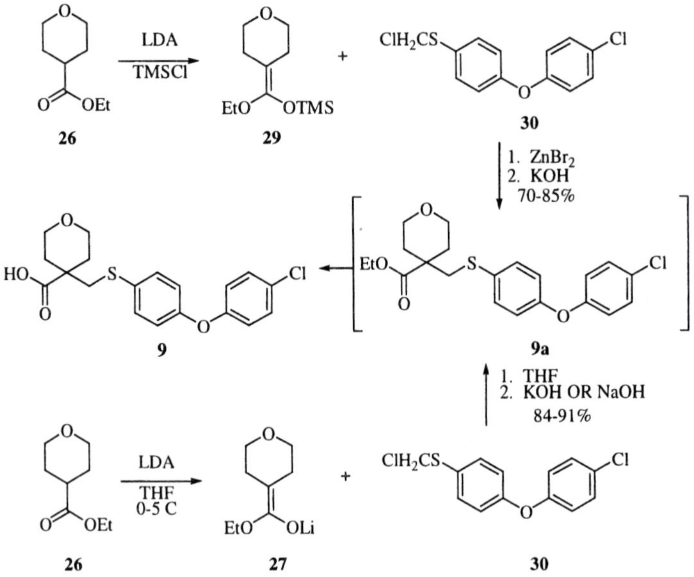

Figure 8: Preferred pyran alkylation routes

Supply of Advanced Intermediates

Of particular concern was supply of advanced intermediates. Large quantities of building blocks of various types were required for route selection, but as the program advanced, it became clear that outsourcing would shorten the "in house" synthesis and decrease cost of goods.

Synthesis of *p*- chloro-*p*'-thiophenyl ether

As the synthesis of **1** evolved, the targeted intermediates changed in detail but remained constant in overall composition: a pyran containing moiety and a *p*-chloro-*p*'-thiophenyl ether based intermediate. Early routes, from the medicinal route to several iterations into the process supply route relied on the general route shown in figure 9. Variations on this synthesis, but adherence to the general sequence while eliminating isolations, distillation and some purifications led to the sequence shown in figure 10.[7] Of concern was isolation of sulfonyl chloride **13**, a highly corrosive, water sensitive solid. Fortuitously, this could be avoided if care was taken to drive the sulfonylation/chlorination step to completion.[8] The thiol produced *via* the route shown in figure 8 was a waxy solid which was used "as is" to open lactone **7** or displace the tosyl group in **25** for synthesis of acid **9** in early process routes. Of note is an alternative to the use of PPh$_3$ for the reduction of sulfonyl chloride **13** to thiol **8** (figure 9).[9,10,11] This is an interesting reaction to scale up because it is quite exothermic. For example, it proved difficult to remove the amount of heat generated by the addition of acetic acid to a suspension of zinc dust in DMF at ambient temperatures, especially if the addition is uncontrolled. However, controlled addition of acetic acid at elevated (ca. 100 °C) temperatures to a suspension of zinc and sulfonyl chloride **13** smoothly afforded thiol **8**.

Clearly, the use of triphenyl phosphine as a reductant presented a cost and waste stream problem. However, for early supply purposes, these routes were sufficient and scaleable.

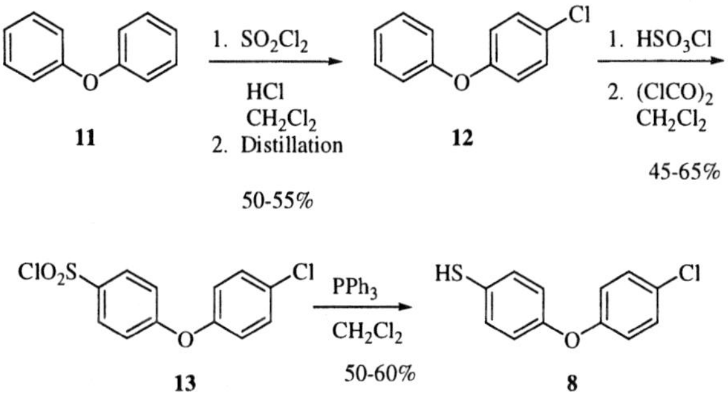

Figure 9: Elaboration of phenyl ether 11 to 8

Thiol **8** was easily converted to methylsulfide **31** *via* the action of various bases and iodomethane. It proved straightforward to chlorinate **31** with sulfuryl chloride to provide a ready source of the chloromethyl substrate **30** for development of the Mukaiyama-type coupling discussed earlier.

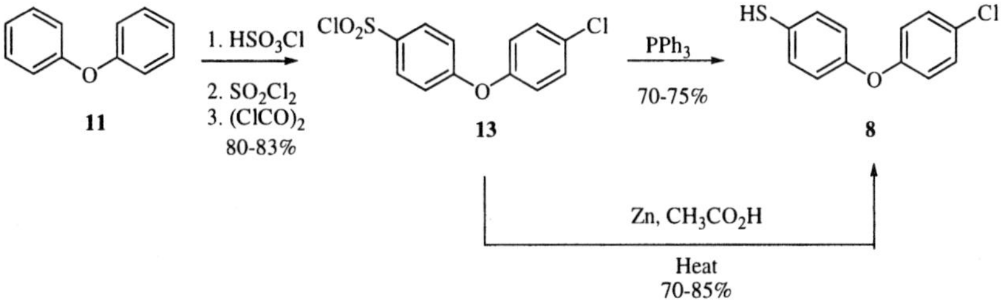

Figure 10: Alternate syntheses of 8

However, the sequence of reactions just described had several drawbacks, not the least of which is the isolation of thiol **8** and subsequent iodomethane mediated methylation. Therefore, an alternative synthesis and methylation procedure was developed, as depicted in figure 11. It was reasoned that an efficient one-pot synthesis of sulfonyl chloride had already been developed and that this could be leveraged to directly produce thiomethyl **31** without the problematic isolation of thiol **8**. This reaction, known as the Michaelis-Arbuzov reaction, proceeds *via* an *O,O,S*-trisubstituted phosphorothioate methyl ester and in this case methyl chloride. Treatment of the reaction mixture with aqueous KOH produces **31** directly, in analogy with findings already published.[7] This proved to be quite useful to our intermediate and larger scale supply campaigns.

Methylsulfide **31** is most easily converted to the chloromethyl intermediate **30** by the action of sulfuryl chloride. This highly reactive intermediate was never isolated but used immediately in the coupling reaction utilising silylketene acetal **29** and ZnBr$_2$ or anion **27** directly.

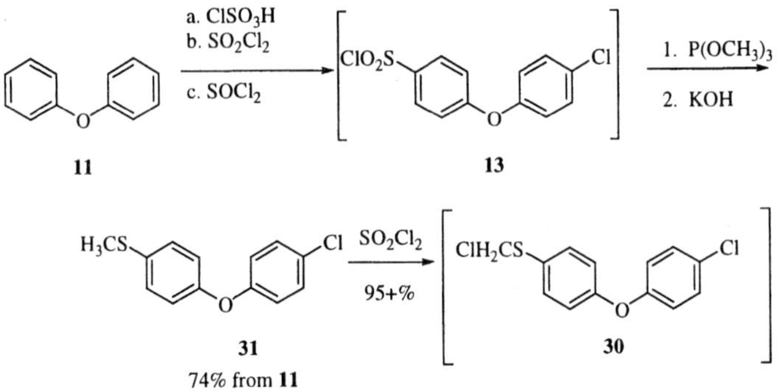

Figure 11: Synthetic route to chloromethyl intermediate 30

Synthesis of Pyran Esters

Pyran di- and monoethyl esters **4** and **26** presented their own particular challenges because they are liquids. Purification by distillation is an obvious yet challenging solution, but was made workable by carefully controlling the impurities arising from the reaction of 2-chloroethyl ether with diethyl malonate. Eventually, direct production of the monoester **26** without isolation of **4** or distillative purification was achieved (figure 12).

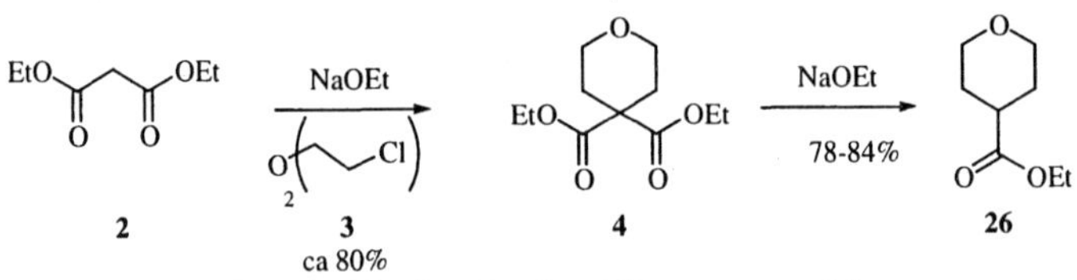

Figure 12: Synthesis of Pyran ester 26

Current Plant Synthesis

Currently, methylsulfide **31** and pyran monoester **26** are obtained from outside suppliers. They are produced using in-house developed syntheses proprietary either to Roche or the suppliers themselves. These advanced intermediates must meet stringent purity and cGMP criteria because all production now is producing clinical API. The synthesis is highly convergent, efficient, and four steps from ester **26** and phenyl ether **31** (figure 13). Yields per step average >90% and are reproducible. The overall yield is 67% from **26** and **31**. Most importantly, API is produced which is 99+% pure with no impurities present in greater than 0.1%, meeting the quality criteria for clinical use.

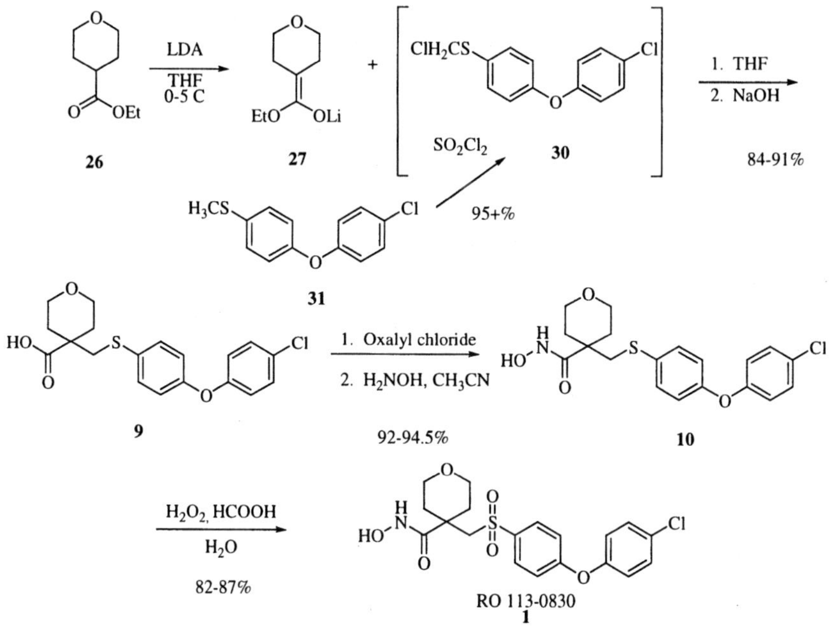

Figure 13: Preferred route to Ro 113-0830

References

(1) Campbell, Jeffrey Allen; Dvorak, Charles Alois; Fisher, Lawrence Emerson; McGrane, Paul Leo. Eur. Pat. Appl. 1999.
(2) Webb, Kevin S. *Tetrahedron Lett.* **1994**, *35*, 3457-60.

(3) Bender, Steven Lee; Broka, Chris Allen; Campbell, Jeffrey Allen; Castelhano, Arlindo Lucas; Fisher, Lawrence Emerson; Hendricks, Robert Than; Sarma, Keshab Eur. Pat. Appl. 1997, 780 386 A1 19 970 625 CAN 127:135724.

(4) (a) Xue, Chu-Biao; Decicco, Carl P.; Wexler, Ruth R. PCT Int. Appl. 1999, WO 9958528 A1 19991118 CAN 131:336951 AN 1999:736703. (b) Robinson, Ralph Pelton PCT Int. Appl. 1999, WO 9 952 910 A1 19 991 021 CAN 131:286406 AN 1999:672818. (c) Montana, John Gary; Baxter, Andrew Douglas; Owen, David Alan; Watson, Robert John PCT Int. Appl. 1999, WO 9 940 080 A1 19 990812 CAN 131:144512 AN 1999:511148.

(5) Maruoka, K.; Yamamoto, H. *Tetrahedron* **1988**, *44*, 5001.

(6) (a) Han, Jeong Sik; Kim, Sang Bum; Mukaiyama, Teruaki *Bull. Korean Chem. Soc.* **1994**, *15*, 529-30. (b) Kobayashi, Shu; Suda, Shinji; Yamada, Masaaki; Mukaiyama, Teruaki *Chem. Lett.* **1994**, *1*, 97-100. (c) Minnow, Nobuto; Mukaiyama, Teruaki *Chem. Lett.* **1987**, *9*, 1719-22.

(7) (a) Cadogen, J. I. G. *Q. Rev. Chem. Soc.* **1962,** *16,* 208 and references cited therein. (b) Klunder, J. M., Sharpless, K. B. *J. Org. Chem.*, **1987**, *52*, 2598 and references cited therein. (c) Mukaiyama, T., Nagaski, U. *Tetrahedron Lett.* **1967**, *35*, 5429.

(8) (a) Abramovitch, Rudolph A.; Azogu, Christopher I.; McMaster, Ian T.; Vanderpool, Danny P. *J. Org. Chem.* **1978**, *43(6)*, 1218-26. (b) Olah, George A.; Ohannesian, Lena; Arvanaghi, Massoud *Synthesis* **1986**, *10*, 868-70. (c) Watson, William David *J. Org. Chem.* **1985**, *50*, 2145-8.

(9) Pfister, Theodor; Schenk, Wolfgang; Blank, Heinz Ulrich Ger. Offen. 1979, DE 2 743 541 19 790 405.

(10) Schenk, Wolfgang; Blank, Heinz Ulrich; Hagedorn, Ferdinand; Evertz, Werner Ger. Offen. 1979, DE 2 743 540 19 790 405.

(11) Blank, Heinz Ulrich Ger. Offen. 1978, DE 2 721 429 19 781 116.

Chapter 7

Four Generations of Pyrrolopyrimidines

Michael F. Lipton[1], Michael A. Mauragis[1], Michael F. Veley[1], Gordon L. Bundy[2], Lee S. Banitt[2], Paul J. Dobrowolski[2], John R. Palmer[2], Theresa M. Schwartz[2], and David C. Zimmerman[2]

[1]Chemical Process Research and Preparations and [2]Combinatorial and Medicinal Chemistry, Pharmacia, 700 Portage Road, Kalamazoo, MI 49001

The pyrrolopyrimidines have been under study for several years at Pharmacia Corporation for a variety of potential therapeutic applications. This class of compound is typified by PNU-142731A (**24a**) and contains a tricylic, generally fully aromatic nucleus, to which are appended two pyrrolidine moieties and a heterocyclic-containing side chain. This paper will present the process development aspects of the synthesis of four candidates in the series, and the application of lessons learned as the clinical program continued to evolve.

Scheme 1

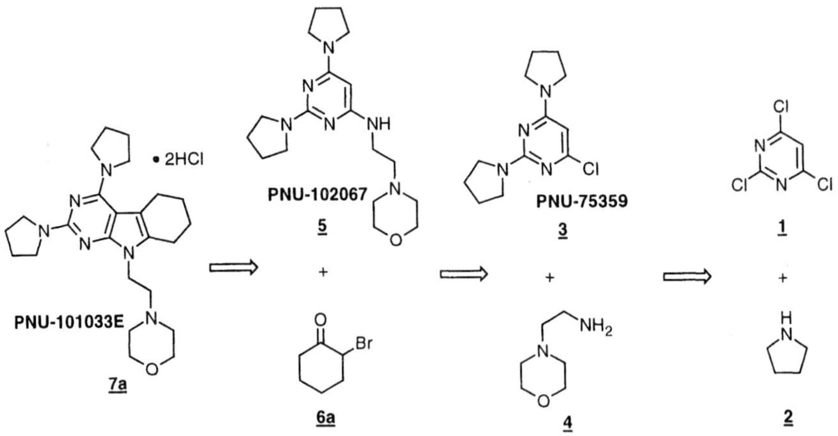

The initial clinical candidate in the series was PNU-101033E (**7a**), which was constructed as depicted retrosynthetically in Scheme 1 (*1,2*). Treatment of 2,4,6-trichloropyrimidine with 2.1 equivalents of pyrrolidine in hot heptane with excess potassium carbonate afforded PNU-75359 in a remarkably selective reaction. The few percent of off isomer generated in this reaction was problematic for earlier workers in a related project. They were forced to develop an acidic work-up procedure to selectively hydrolyze the undesired isomer at the expense of only ca. 10% of PNU-75359. As we later determined, the off isomer was rejected in the cyclocondensation reaction of this sequence and the purification of PNU-75359 was unnecessary. The >19:1 mixture can be simply carried through the sequence as is. This simple change increased the yield from 82.4% to >95%, and significantly simplified the process. On modest scale, displacement of the remaining chloride in the now deactivated system was accomplished by heating with excess aminoethylmorpholine. Initially, the expensive aminoethylmorpholine was used as solvent for this transformation since it will unfortunately not proceed at synthetically useful rates at temperatures below ca. 180°C. The preparation of laboratory quantities by this method was acceptable, but the inconvenient temperature, dangerous exotherm and isolation difficulties of the cooled solid reaction mass made the development of an alternative scalable method a priority. Although it was felt that dilution with a high boiling solvent would solve the safety, handling, and cost issues, this in fact proved not to be the case. In fact, even with added auxiliary bases the displacement could not be made synthetically viable. We reasoned that if we increased the nucleophilicity of the morpholine reagent, perhaps more reasonable large scale reaction conditions could be successful. Thus treatment of aminoethylmorpholine with an equivalent of n-butyllithium resulted in the generation of lithium amide base **8**. This species was sufficiently nucleophilic to displace the chloride in our deactivated substrate at convenient production temperatures. In the process procedure, aminoethylmorpholine was slurried in heptane at –25°C and treated with an equivalent of n-butyllithium. This results in a gentle, easily controllable exotherm as the highly insoluble amide base **8** was generated. A heptane solution of PNU-75359 was then added and the reaction mixture heated to a gentle reflux for 18 hours. The solid product and reaction salts were then filtered and dissolved off the filter with methylene chloride and water. An aqueous partition afforded the trisubstituted product **5** in near quantitative recovery contaminated only with a minor amount of off isomer from the initial pyrrolidine displacement.

Scheme 2

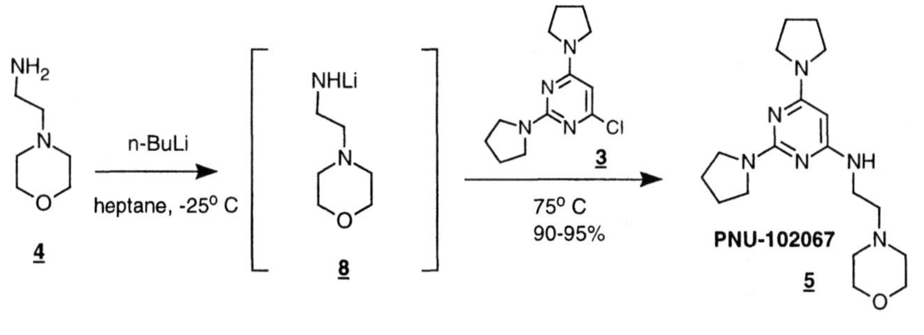

The complete molecular framework is then constructed via a one pot Bischler type alkylation/cyclization/dehydration reaction (*3*) with 2-bromocyclohexanone. To conduct this chemistry on scale, we required large quantities of 2-bromocyclohexanone. Due no doubt to its limited thermal stability, it is not commercially available. It can be prepared by a simple bromination of cyclohexanone with Br_2 in methanol, but this results in significant overreaction and the thermally labile product must be purified by vacuum distillation. We required a method that would avoid the distillation of this hazardous material and developed a two step protocol via the TMS enol ether. It is both high yielding and avoids the safety issue. Thus treatment of cyclohexanone in hot dimethylformamide with triethylamine and trimethylsilyl chloride resulted in smooth conversion to the TMS enol ether. After an aqueous partition with hexane, the enol ether was then converted to 2-bromocyclohexanone by the action of 0.98 equivalents of N-bromosuccinamide. The enol ether intermediate may be used as is for the conversion after an aqueous partition. However, we generally chose to vacuum distill it prior to conversion to 2-bromocyclohexanone since stoichiometry in the bromination step proved critical. It should be noted that we experienced a chemical instability upon attmpting to dissolve N-bromosuccinimide in THF solvent at concentrations in the 10% weight/volume range. For this reason, the NBS had to be portion-wise added as a solid to the cooled substrate for best results.

With large quantities of clean 2-bromocyclohexanone in hand, we next turned our attention to the Bischler type reaction. Our initial attempts at conducting this reaction were disappointing. Yields were only in the 43-60% range and extensive purification of the PNU-101033 free base was required. The chemistry was actually remarkably clean, but simply incomplete with almost the entire mass balance being made up of either starting material or desired product. We surmised that the reaction could be equilibrium controlled, and felt that the opportunity existed for driving the equilibrium by removing the water generated during the dehydration. This proved to be the case. When two and one-half equivalents each of diisopropylethyl amine and 2-bromocylohexanone were added incrementally to a reaction mixture gently refluxing through a bed of 4A molecular sieves, a >90% conversion to PNU-101033 was realized. An added bonus of this procedure was that the product was now pure enough to crystallize directly from the cooled reaction mixture, and therefore no work-up was required. Of course such a procedure cannot be scaled directly into process equipment, but by taking advantage of the water:acetonitrile azeotrope and conducting the reaction in a gentle distillative mode, the same results could be obtained on scale.

Reproducible conversion to a pharmaceutically acceptable pure salt proved not to be trivial in the present case. In fact, despite their obvious structural similarities, all of the compounds studied in this series required the development of their own salt formation procedure. They all required different conditions for salt generation due to the propensity of the pyrrolopyrimidines to retain solvents and moisture, in addition to solubility concerns. We finally settled on treatment with two equivalents of HCl in a mixed solvent system of 9:1 ethyl acetate:methanol. This choice of solvent mixture allowed for the warm dissolution and polishing of the sparingly soluble free base and a high recovery of >99% pure solvent-free final product. The overall chemical sequence to produce PNU-101033E in the forward sense is shown in Scheme 3.

Scheme 3

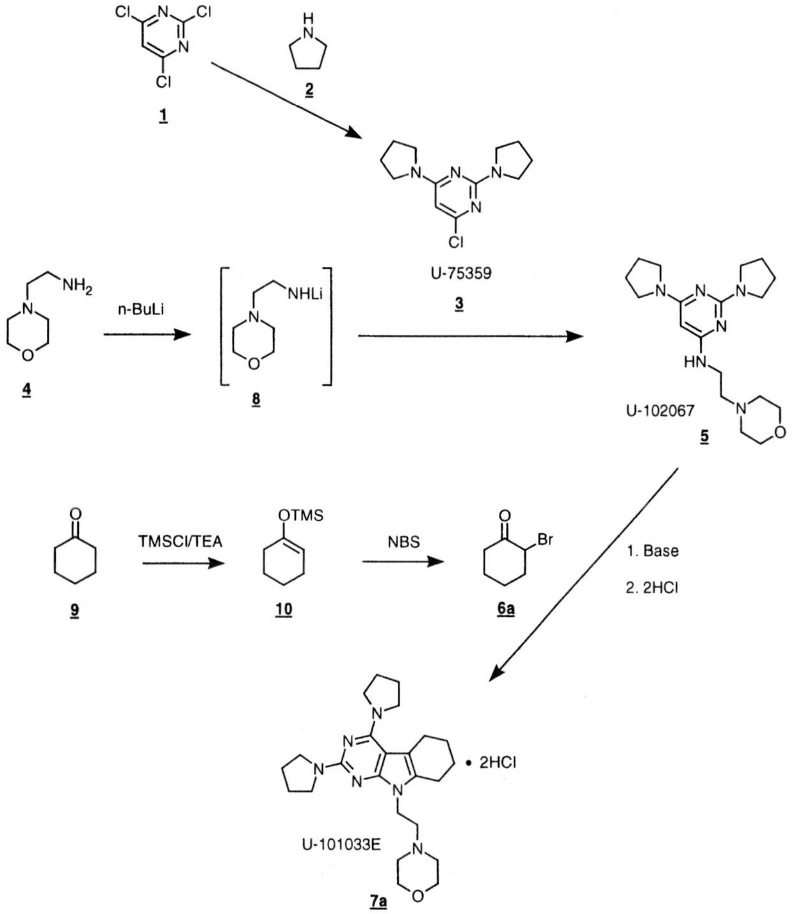

Unfortunately, development of PNU-101033E ended as a result of an unusual dose dependant liver toxicity uncovered during the 90 day path tox studies. However, our Medicinal Chemistry colleagues promptly proposed the replacement candidate PNU-104067. This material should be readily available along the same synthetic pathway via an oxidation of the 6-membered carbocyclic ring. Such an oxidation is handily carried out on laboratory scale using excess DDQ or chloranil. However, these oxidants are invariably used in excess, the reactions rarely go to completion, and a tedious chromatographic purification is generally required to obtain pure product. With the known liver toxicity of PNU-101033, the level of residual starting material in clinical lots of PNU-104067 product could not exceed 0.5%, further dooming this approach. We viewed this oxidation reaction as simply the reverse of a transfer hydrogenation, i.e., in a transfer hydrogenation a suitable donor is used to reduce the acceptor. In the present case, the donor molecule would in fact be the substrate, leading to its oxidation rather than reduction. However, under standard phase transfer conditions, the reaction could not be pushed to completion. In fact, it was difficult to push the reaction beyond about 50% completion in an autoclave or bomb employing a wide variety of acceptors, catalysts, and additives. The rate of oxidation was quite manageable, however, and as we later found, dependant on substrate quality. We determined that it was possible in the lab to simply run the reactions at reflux in a suitable high boiling solvent under a nitrogen bubbler, allowing the hydrogen generated to dissipate (Scheme 4). In this fashion, the starting material was consumed completely in 12-24 hours and replaced quantitatively by near-analytical quality PNU-104067, **11**. Work-up consisted of methylene chloride dilution followed by filtering off the catalyst using solka floc on a Sparkler filter and washing the filter cake with methylene chloride. A concentration in vaccuo to remove the methylene chloride left clean PNU-104067. The product precipitated from the non-polar high boiling solvent in near quantitative yield in >98% purity.

Scheme 4

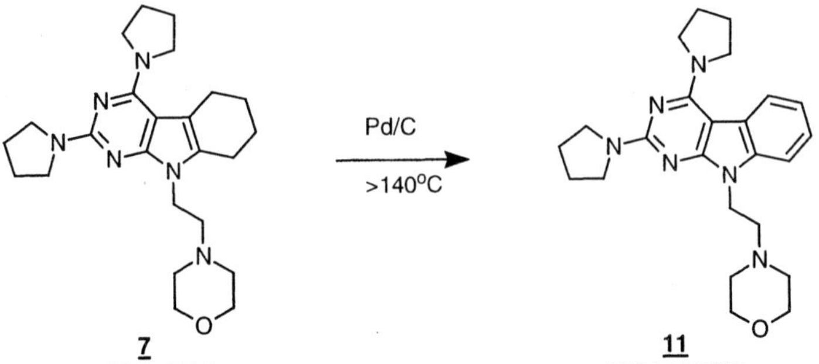

7
PNU-101033

11
PNU-104067

Conversion to the monohydrochloride hemihydrate was accomplished by treatment with one equivalent of HCl in 95% ethanol. In this case an ethyl acetate/alcohol mixture provided an uncrackable ethyl acetate solvate. Unfortunately, disaster struck again in the clinical program as although the toxicity issue had been adequately addressed with the structural modification, PNU-104067 exhibited metabolic instability in man. The major human metabolite in circulating blood was the side chain oxidized PNU-141571. Although known, interestingly this material was but a minor metabolite in the animal studies.

Our synthetic target therefore turned briefly to PNU-141571, the synthesis of which is shown in Scheme 5. It begins with a displacement on the deactivated monochloropyrimidine **3** with ethanolamine. Safety studies had confirmed our suspicions about the thermal hazards associated with this displacement, hazards typical for displacements in this series. Moreover, activation as the amide base was unsuccessful in the present case. Attempted displacement with either the monoanion or the presumed dianion (generated by addition of one or two equivalents of n-butyllithium to ethanolamine) afforded the ether exclusively. We were likewise unsuccessful in our attempts to selectively protect the alcohol function of ethanolamine, and had to resort to multiple large-scale lab runs in order to stockpile the requisite trisubstituted product needed to produce material for preclinical studies. Despite the 82% yield for this transformation, our inability to scale this step beyond large lab scale was problematic. The Bischler-type cyclization, although carried out as previously and to a similar conversion, failed to produce spontaneously crystalline product in this series and we eventually had to resort to a chromatographic purification of the tricycle in order to proceed. It was obtained in 76% yield. Displacement with an additional mole of ethanolamine, activation and acylation, happily, were uneventful. The acylation occurred selectively even at the more process friendly $-30°C$ rather than $-78°C$. We accomplished the cyclization with KOt-Bu in THF, a standard Process tool, rather than the often-employed Discovery NaH procedure. The penultimate was generated in an acceptable 38% overall yield from amino alcohol **14**. The oxidation was conducted as in the PNU-104067 case, utilizing Pd/C at 143°C, obtaining a near quantitative yield of free base. The hydrochloride salt was cleanly generated in 87% yield employing acetone/methanol as the solvent mixture with a subsequent concentration prior to product isolation, as the procedures used successfully in the previous candidates both failed to afford a clean, solvent-free salt in acceptable recovery.

Scheme 5

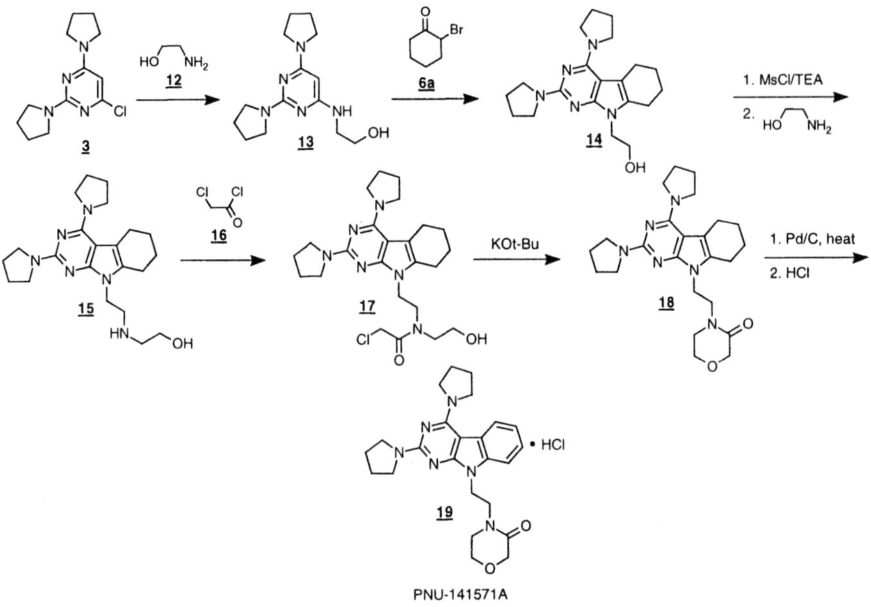

PNU-141571A

With the problematic synthesis of PNU-141571A, we were more than happy to have it wash out of the clinical program, again because of toxicity issues. The Medicinal group was once again immediately ready with a back-up candidate, PNU-142731. This final candidate in the series was originally synthesized from stockpiled supplies of chromatographed ethanolamine adduct **13** left over after the demise of the PNU-141571 project. It was known from the outset, however, that if quantities larger than a few kilograms of PNU-142731 were to be required, an entirely new synthesis would have to be developed. The so-called A-Process (chronologically the initial synthetic route) to produce kg quantities of PNU-142731A is depicted in Scheme 6. It should be noted that with the close cooperation of Medicinal and Process chemists, this route was developed and used to prepare over 6 kg of clinical quality material while the B-Process route, depicted in Scheme 7, was both developed and demonstrated on large lab scale.

Scheme 6

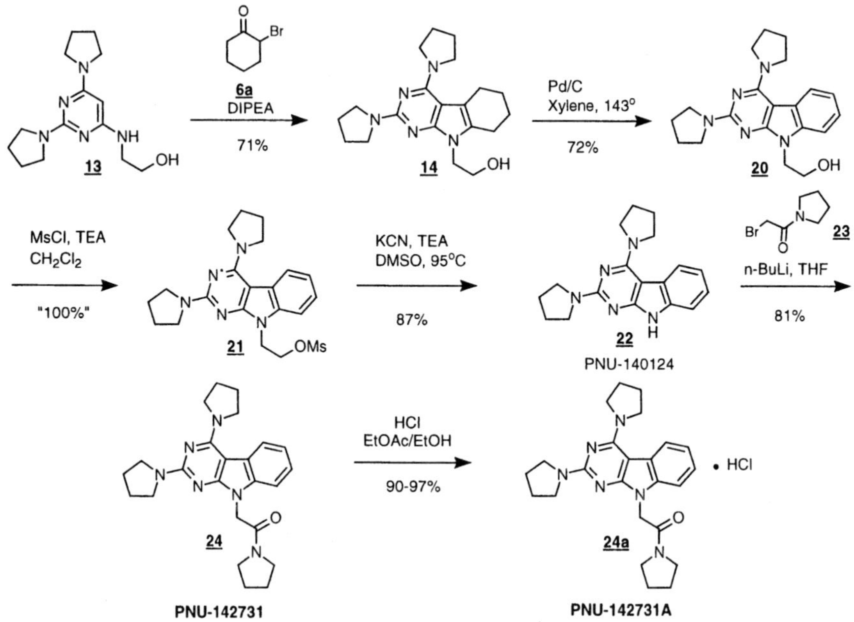

The tricycle is constructed in the usual fashion with the now familiar catalytic oxidation of the six membered ring occurring earlier in the sequence. The side chain is next installed, but via a rather circuitous pathway which requires some explanation. The original intention was to oxidize the terminal carbon of the side chain to the carboxylic acid and convert to the amide. However, all attempts at this oxidation, including TEMPO, K_2CO_3/I_2, buffered $KMnO_4$, activated MnO_2 and peracid were unsuccessful even on small lab scale, providing yields of 50% or less. It was possible to oxidize cleanly and in high yield to the aldehyde using Swern conditions (4), but this compound could not be cleanly oxidized further in acceptable yield, despite attempts with a variety of reagents. In hindsight, this should not be very surprising, since the pyrrolopyrimidines grew out of an anti-oxidant program. The alcohol was therefore activated and subjected to a retro Michael reaction by the action of excess potassium cyanide. Although cyanide use in dedicated process equipment poses little risk and essentially requires no extra processing operations, its use in batch processes in multipurpose equipment such as our pilot plant is problematic from a waste generation and clean up standpoint. Additionally, early tox data indicated that the product of the retro Michael, PNU-140124, was Ames positive (5) which would seriously complicate handling. For these reasons, we elected to contract this and the ensuing step out to a toll conversion house even though the chemistry was being performed on what would constitute only a modest pilot plant scale. The PNU-142731 was returned to us and we developed yet another unique, reproducible, and high yielding procedure for the generation of clinical quality hydrochloride salt.

Having material toll converted in the middle of the sequence gave us the opportunity to work with our Medicinal Chemistry colleagues and develop what became the route to 100 kg quantities of drug, the B-Process, depicted in Scheme 7.

Scheme 7

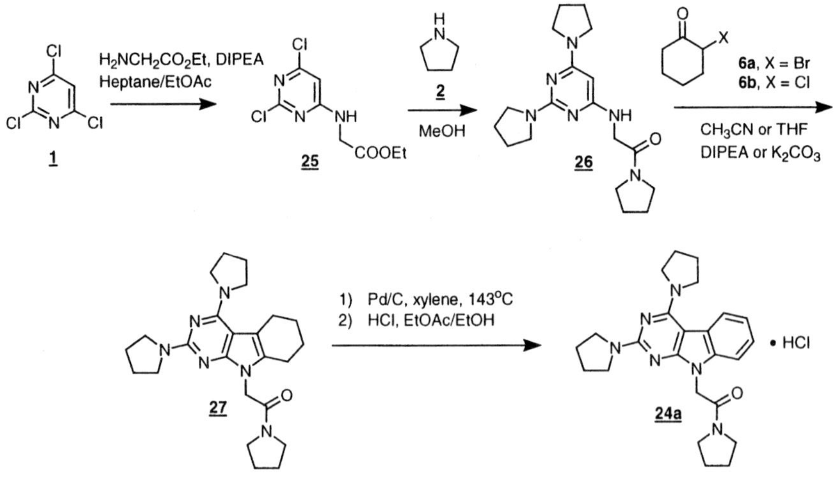

This route is short, efficient, and the chemistry is robust at any scale. The keys of the process are the required reordering of the chloride displacements and the operationally complex double displacement/pyrrolidine amidolysis. There is obviously a statistical bias for substitution at the 4- and 6-positions of the activated trichloropyrimidine. It is possible to enhance this bias even further by conducting the displacement at or below room temperature using only a slight excess of glycine ethyl ester hydrochloride. In fact, we were able to reproducibly obtain on scale a usable 4:1 ratio of the desired regioisomer, a ratio which is more tolerable as the first step of the sequence than it might be later on. The solvent choice was not critical as we determined that methanol also provided a similar ratio of products, but heptane/ethyl acetate allowed a facile isolation of product in 72-76% yield via heptane dilution and knock out. The product was invariably contaminated with a small percentage of off-isomer, but as in the PNU-101033 and PNU-104067 cases above, this posed no problem since it was handily rejected later in the sequence.

The displacement and amidolysis required some operational gyrations to be run successfully. Displacement of the first of the ring chlorides was very exothermic, while displacement of the second required gentle heating. In our initial experiments, we isolated and characterized both the mono- and bis-pyrrolidinylpyrimidine intermediates. In order to obtain reference samples of the amide final product, the bis-substituted material was then carried on by treatment with pyrrolidine/ethoxide under forcing conditions. On scale, the safest way of conducting the initial exothermic displacement was with external cooling, employing excess pyrrolidine and using methanol solvent as a heat sink. When this first displacement was complete, the reaction mixture could be heated to reflux to effect the second displacement. The amidolysis, however, required extremely forcing conditions. We were never able to successfully conduct this transformation in the presence of solvent using simply pyrrolidine as the nucleophile. We did determine, though, that the desired reaction did occur in neat pyrrolidine, and folded this transformation into a one-pot sequence. Thus, after the chloride displacements were complete, and methanol had served its purpose, it could be removed by atmospheric distillation, leaving a pyrrolidine solution of the ester. A 16 hour reflux followed by aqueous partition provided a recrystallized 94% yield of the desired amide.

The tricycle is again constructed via the familiar Bischler type cyclization reaction. One may use 2-bromocyclohexanone or the more stable commercially available 2-chlorocyclohexanone as the substrate. An interesting and valuable discovery was that potassium carbonate may be substituted for the diisopropylethyl amine base in this step. Carbonate, in addition to functioning as base, also functioned as dehydrating agent, thus minimizing the effect of slight process deviations during the azeotropic water removal required of the DIPEA procedure. The familiar catalytic dehydrogenation reaction completed the abbreviated sequence to PNU-142731 free base. It was then converted as in the A-Process to clinical quality supplies of the hydrochloride salt.

Synthetic procedures used for the preparation of multi-kilogram quantities of a series of pyrrolopyrimidines, under study as potential oral anti-asthma medications, were described. The key tri-amino substituted pyrimidines were constructed via a variety of procedures. Unfortunately, the most convenient procedure on scale, displacement with a lithium amide base, was found not to be general. The sequences all continued with a key Bischler type cyclization/condensation/dehydration reaction. By removal of the water generated during the course of this reaction, the chemistry was pushed to near completion and proceeded in overall 73-86% yield on scales ranging up to 140 kg input for the various candidates. When development of the initial candidate, PNU-101033E, was abandoned due to toxicity concerns, a facile high yielding catalytic oxidative dehydrogenation procedure was developed to prepare PNU-104067, the second candidate in the series. This procedure proved valuable in the synthesis of the remaining candidates, as well. Although preparation of the clinically short-lived major human metabolite, PNU-141571A, proved problematic on scale, a short, very workable synthesis of PNU-142731, the final clinical candidate, was developed. The successful synthesis required a re-ordering of the installation of the amino functionalities. This final process allowed for the construction of several hundred kilograms of the stockpiled free base and over 100 kilograms of clinical quality monohydrochloride salt. The overall yield of PNU-142731A prepared by the B-Process from 2,4,6-trichlorpyrimidine was in the 42-46% range in 3000 l - 4000 l equipment.

References

(1) Bundy, G.L.; Ayer,D.E.; Banitt, L.S.; Belonga, K.A.; Mizak, S.A.; Palmer, J.R.; Tustin, J.M.; Chin, J.E.; Hall, E.D.; Linseman, K.L.; Richards, I.M.; Scherch, H.M.; Sun, F.F.; Yonkers, P.A.; Larson, P.G.; Lin, J.M.; Padbury, G.E.; Aron, C.S.; Mayo, J.K. *J. Med. Chem.* **1995**, 38, 4161.

(2) Mauragis, M.A.; Veley, M.F.; Lipton, M.F. *Org.Proc.Res.Dev.* **1997**, *1*, 39.

(3) Ayer, D.E. PTC International Patent Application WO9106542A1. *Chem. Abstr.* **1989**, *115*, 114546.

(4) Mancuso, A.J.; Swern, D. *Synthesis* **1981**, 165.

(5) Ames, B.N.; McCann, J.; Yamasaki, E. *Mutation Res.* **1975**, *31*, 347.

Chapter 8

Synthesis, Discovery, and Separation of the Atropisomers of CP-465,022

Keith M. DeVries

Chemical R&D, Pfizer Inc., Eastern Point Road, Groton, CT 06340

The total synthesis of CP-465,022 is described. This former drug candidate exists as two separable rotational isomers (atropisomers). The discovery and separation of these rotational isomers is described, as well as the methodology used to prepare the pyridyl sidechain and the quinazolinone nucleus. Testing showed that the desired biological activity existed in just one of the atropisomers.

Scheme 1

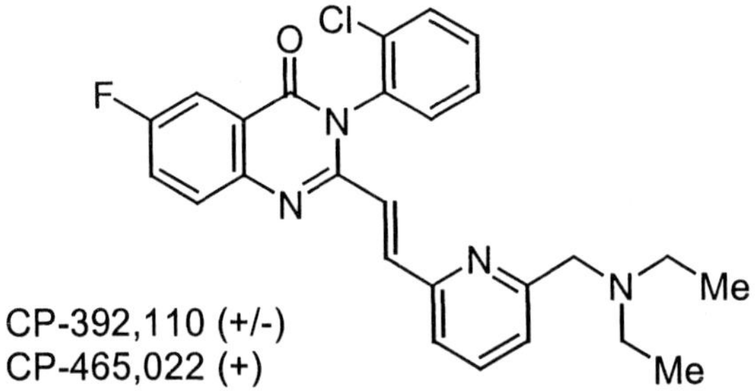

CP-392,110 (+/-)
CP-465,022 (+)

CP-392,110 (racemic CP-465,022) was the lead structure in a Discovery program for the treatment of stroke, with the AMPA (alpha-amino-3-hydroxy-5-methyl-4-isoxazole-propionic acid) receptor being the desired target (Scheme 1). A high throughput screen produced a number of hits in the quinazolinone series. Pfizer's proprietary niche in this area was the aminomethyl-pyridine moiety, which imparted the AMPA antagonism to this series of compounds *(1)*. As noted in the title, CP-392,110 actually exists as two rotational isomers (atropisomers) about the nitrogen – *ortho*-chlorophenyl bond. Rotational isomers which are stable enough to be isolated are referred to as atropisomers. It was determined that all of the AMPA activity resided in one of the enantiomeric rotational isomers, CP-465,022 (*vide infra*).

Scheme 2

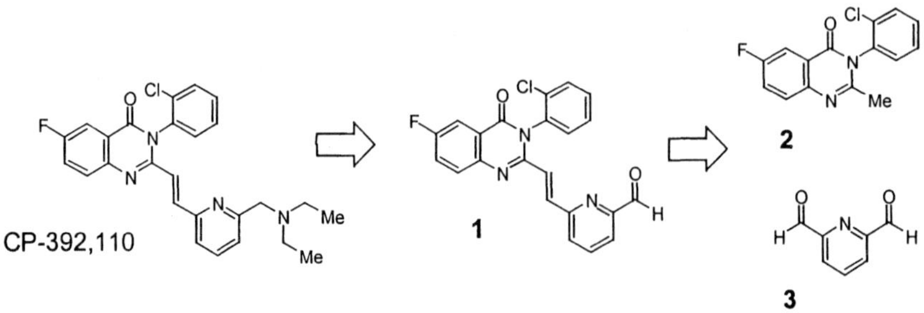

CP-392,110

The Discovery retrosynthesis is shown in Scheme 2, and it was deliberately linear for the purposes of their SAR related to the aldehyde **1**. Once the

Scheme 3

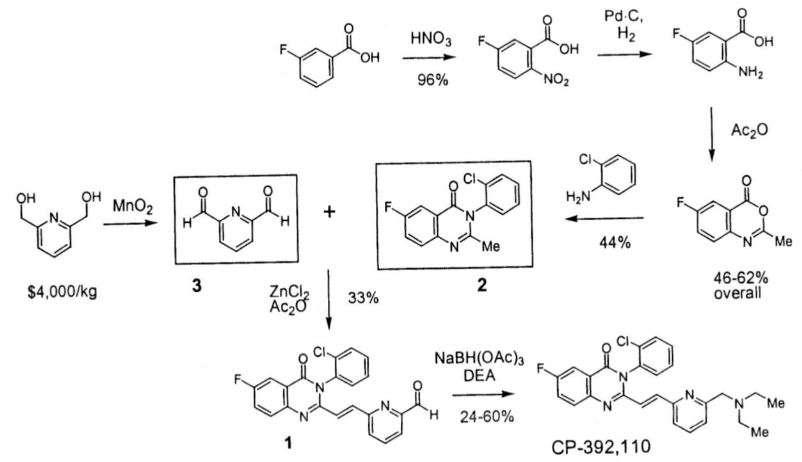

Scheme 4

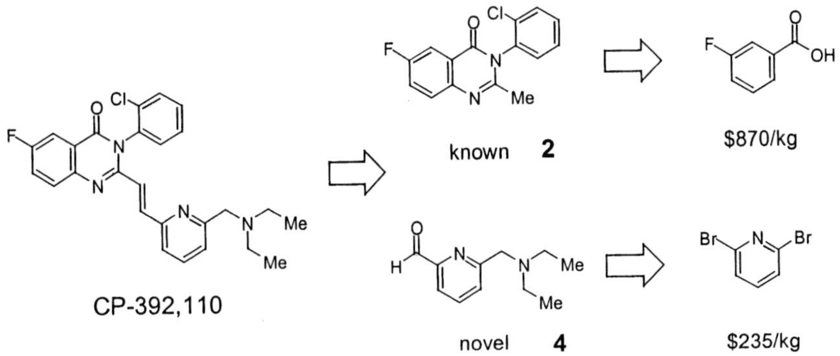

candidate was nominated, this retrosynthesis included one known intermediate, the quinazolinone nucleus **2**, and the novel dialdehyde **3**, which was very difficult to prepare and isolate.

The Discovery synthesis shown in Scheme 3 was linear by design and, therefore, less efficient for larger scale manufacture. As is always the case, however, the Process Research and Development group benefitted greatly from this previous work. The synthesis of the quinazolinone core **2** will be described in more detail later. The major issues with the Discovery route from a scale-up perspective were the preparation and isolation of the dialdehyde **3** and the subsequent aldol addition/elimination to form **1**, which had the potential for "dimer" formation by the reaction of **1** with a second molecule of **2**. Finally, a reductive amination of **1** with diethyl amine provided CP-392,110 (racemic CP-465,022), albeit with variable yields.

Once we knew that CP-392,110 was the final candidate, a more efficient bond disconnection was the one across the olefin linker to provide the same quinazolinone nucleus **2** and a novel aldehyde **4** (Scheme 4). The initial starting materials for each of these were *meta*-fluorobenzoic acid and 2,6-dibromopyridine, which were both available in kilogram quantities at a reasonable price. This disconnection would provide a more efficient synthetic sequence and avoid the potential for "dimer" formation in the aldol addition. The major uncertainty with this approach was finding an efficient synthesis of the aldehyde **4**.

Scheme 5

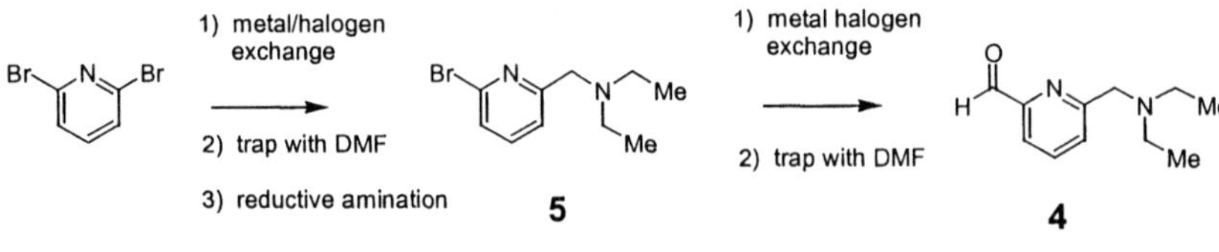

Our plan as outlined in Scheme 5 was to treat 2,6-dibromopyridine with butyllithium, to quench the mono-anion with DMF (dimethylformamide), and then to do a reductive amination with diethylamine to provide the diethylamine **5**. This would be followed by a second metal-halogen exchange and the resultant anion would be trapped with DMF to provide the desired aldehyde **4**. This aldehyde would then be condensed with the quinazolinone nucleus **2** under conditions analogous to those developed by our Discovery group.

Scheme 6

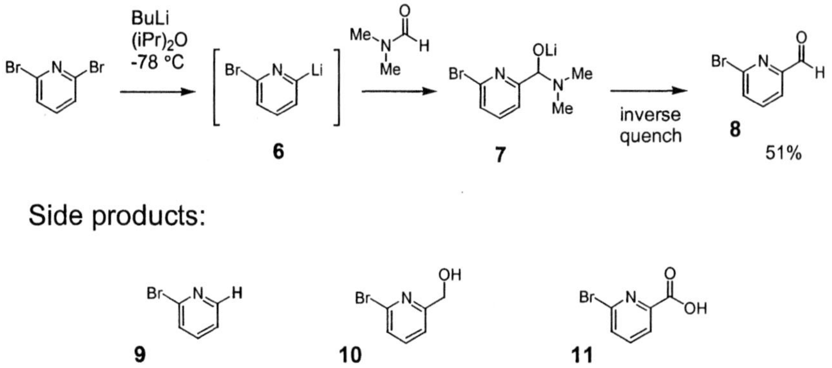

Side products:

    As shown in Scheme 6, 2,6-dibromopyridine was treated with n-butyllithium at -78 °C to form the lithio species **6** which was trapped with DMF (dimethylformamide). The tetrahedral intermediate **7** was then inversely quenched into acid to form the aldehyde **8** in 51% yield. The importance of diisopropyl ether as solvent for the n-butyllithium reaction will be described in detail later. Side products formed in this reaction were the protonated

Scheme 7

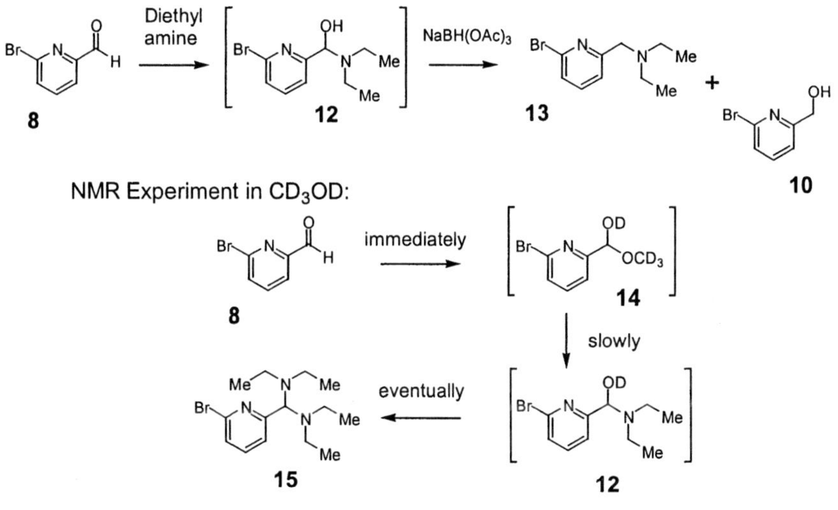

NMR Experiment in CD₃OD:

compound **9** as well as the Cannizzaro products **10** and **11**. The inverse quench was essential to minimize the extent of Cannizzaro reaction.

The aldehyde **8** (Scheme 7) was then treated with diethylamine to form the tetrahedral adduct **12**, which was reduced with sodium triacetoxyborohydride to provide the desired amine **13** and the benzyl alcohol **10**. Using alcoholic solvents, we were never able to keep the reduction product **10** to an acceptable level. In order to study this reaction further, we examined aldehyde **8** and an excess of diethylamine in CD$_3$OD by NMR. Almost immediately, the hemiacetal **14** was formed, followed by slower exchange with diethylamine to form **12**, which eventually reacted with a second equivalent of diethyl amine to form **15**. This suggested that alcoholic solvents should be avoided, as hemiacetal **14** was forming **10** upon treatment with the reducing agent. Indeed, by switching to THF as solvent, the formation of alcohol **10** was almost completely suppressed.

Scheme 8

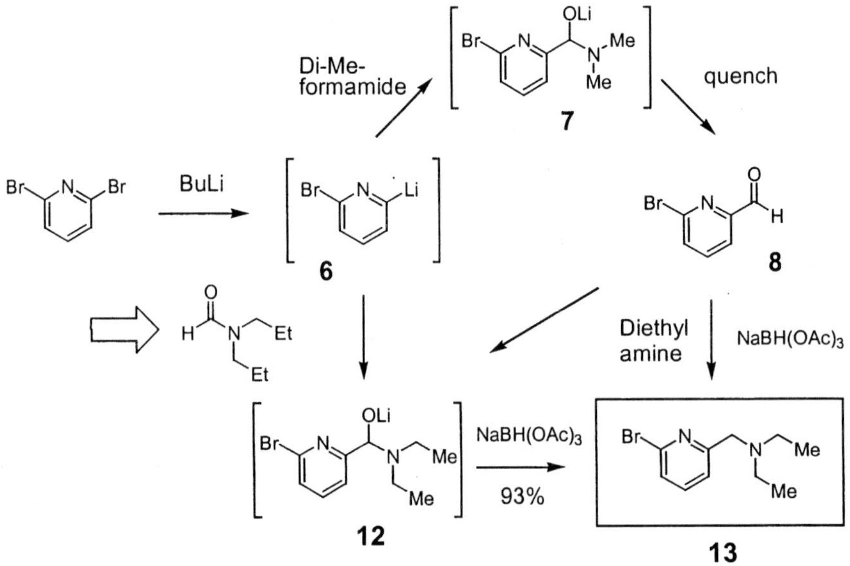

The tetrahedral intermediate **7** formed by quenching the aryllithium **6** with dimethylformamide was directly analogous to the one formed by treating the aldehyde **8** with diethyl amine to form **12** (Scheme 8). This led us to consider the option of trapping the aryllithium **6** with di*ethyl*formamide. Reduction would then provide the desired benzylic amine **13** without proceeding through the aldehyde **8**. In the event, trapping of the bromolithium species **6** with diethylformamide to form **12** followed by direct reduction with sodium triacetoxyborohydride provided the desired benzylic amine **13** in 93% assayed yield (the product is an oil). This material was of sufficient purity to carry into the next reaction directly.

Scheme 9

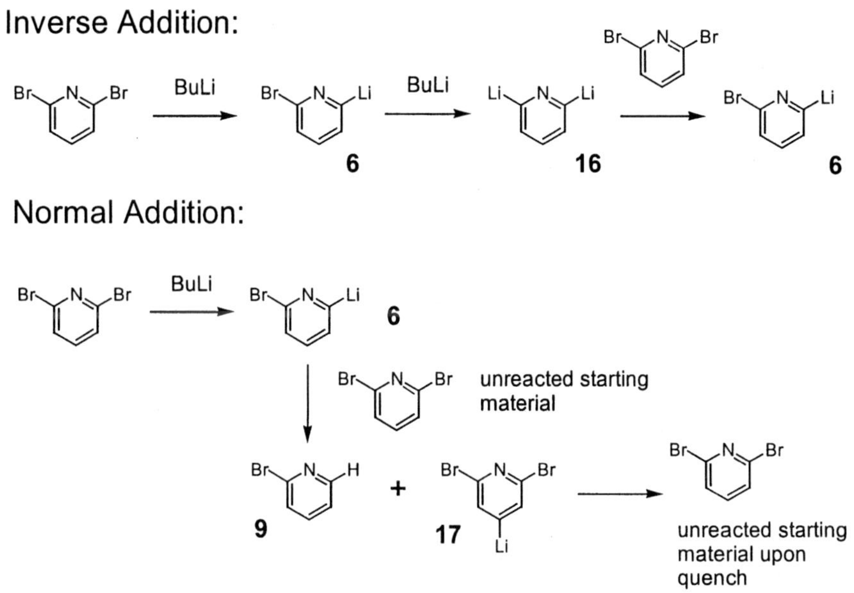

Inverse Addition:

Normal Addition:

While we were engaged in this work, a report from our Process colleagues at Merck was published (*2*). In this report, they described the treatment of 2,6-dibromopyridine with butyllithium in THF as solvent and that inverse addition of starting material to butyllithium was critical for success (Scheme 9). With an inverse addition, they described that the dilithiated species **16** forms and disproportionates to the monolithio species **6** during the second half of the addition of 2,6-dibromopyridine. With a normal addition, they described that the aryllithium species **6** was basic enough to deprotonate unreacted starting material to form lithiated species **17**, along with the dehalogenated starting material **9**. Upon quenching, the lithio species **17** was protonated to reform starting material. We also observed that some reactions did not appear to go to completion, even upon addition of excess butyllithium. The control experiments carried out by the Merck group provided an excellent explanation for these results. With regard to our use of diisopropyl ether as solvent, the 2,6-

dibromopyridine starting material was essentially insoluble in diisopropyl ether at -78 °C, thus effectively mimicking the inverse addition protocol developed by the Merck group.

There is considerable literature precedent with respect to the metallation of 2,6-dibromopyridine to form **6**. The first report was from Gilman with THF as solvent, which described a rapid addition of the butyllithium in order to have an efficient reaction (*3*). Although feasible on smaller scale, this is not viable on multi-kilogram scale. This was followed by the report from Holm which utilized diethylether (*4*). Holm reported that diethylether was a better solvent than THF, as the starting material was insoluble in this solvent, effectively mimicking the inverse addition developed by the Merck group. This report describes the use of diisopropyl ether as a more process friendly solvent than diethylether. Finally, a report from Peterson described the use of methylene chloride as the reaction solvent (*5*). With methylene chloride, the ring deprotonation was controlled as well.

Scheme 10

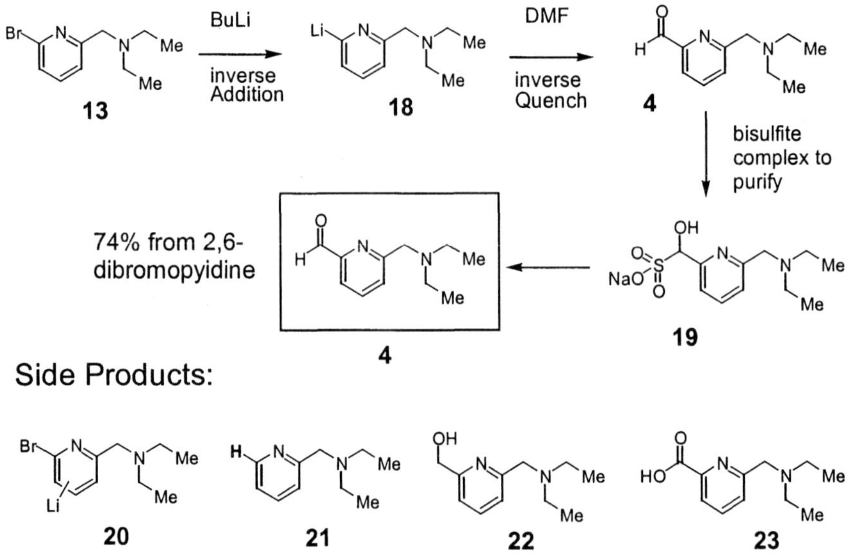

Side Products:

The final reaction sequence to prepare the aldehyde **4** is shown in Scheme 10. The second metal-halogen exchange to form **18**, with an inverse addition of **13** to butyllithium in analogy to the Merck report, was followed by addition of DMF to trap the aryllithium. An inverse quench of this tetrahedral intermediate into aqueous acid provided the desired aldehyde **4**. In order to purify this material, the water-soluble bisulfite addition complex **19** was formed. The desired compound **4** was isolated as an oil in 74% overall yield from 2,6-dibromopyridine and was of sufficient purity (~95%) to use in the final aldol addition reaction to form CP-392,110. A number of minor side products were present: unreacted starting material from ring deprotonation (minimized as a result of the inverse addition of **13** to butyllithium), some protonated starting material **21**, and the benzyl alcohol **22** and benzoic acid **23** derived from the Cannizzaro reaction (these were minimized with the inverse quench).

Control of pH during the bisulfite complex formation was critical. The pH during the formation of the complex **19** was slightly acidic. If organic washes to purge non-aldehydic impurities were performed at this pH many of the impurities remained in the aqueous layer, due to the basicity of the amine. By adjusting to pH 8.5, however, the non-bisulfite adducts partitioned into the organic layer. It was important to not have the pH be too high at this stage or the bisulfite complex **19** was broken to reform **4**. Finally, by adjusting the pH to 10.5, the bisulfite complex was broken and the neutral aldehyde **4** partitioned into the organic layer. If the pH was above 10.5, however, the aldehyde itself would hydrate and remain in the aqueous layer.

Scheme 11

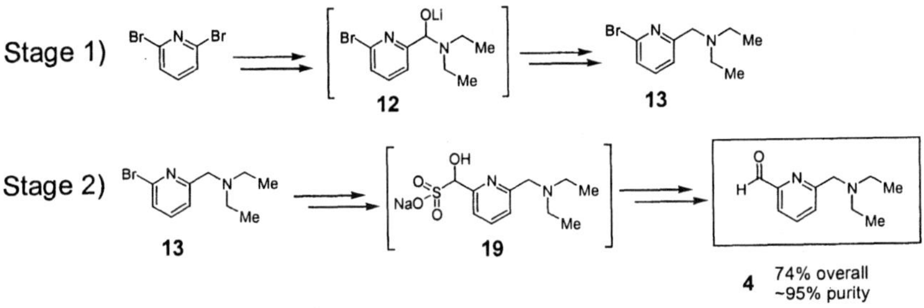

The combined sequence to aldehyde **4** is shown in Scheme 11. 3.0 kilograms of 2,6-dibromopyridine was converted to 1.8 kg of **4** in 74% overall yield. The trapping of the initial aryllithium **6** with di*ethyl*formamide to form **12** and direct reduction of this tetrahedral intermediate allowed us to form the benzylic amine **13** directly without proceeding through the aldehyde **8**. Diisopropyl ether was utilized a reaction solvent, which mimicked the inverse addition protocol reported by the Merck workers. After the second metal halogen exchange with **13** and trapping with DMF, the desired aldehyde **4** was purified as the bisulfite complex **19**, which required careful adjustment and control of pH. Finally, it is worth noting that product **4**, of molecular weight <200 and with no chiral centers, was far and away the most challenging aspect of this project!

Scheme 12

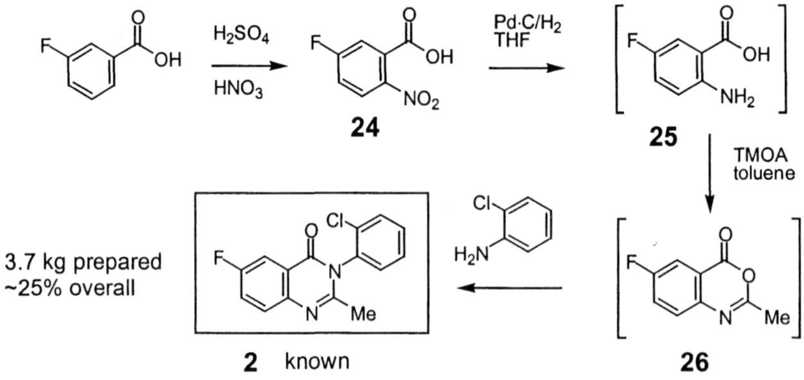

With the synthesis of the sidechain **4** in hand, we turned our attention to the quinazolinone nucleus **2**, which is a known compound. We carried out a reaction sequence which was a modification of the one in the literature (*6*). As shown in Scheme 12, *meta*-fluorobenzoic acid was nitrated to form **24**. This reaction will be discussed in more detail (*vide infra*). The nitro group in **24** was reduced with Pd·C/H$_2$ to form **25**, and the benzoxazinone **26** was formed by treatment of **25** with trimethylorthoacete (TMOA). Treatment of compound **26** with *ortho*-chloroaniline provided the quinazolinone **2**. This sequence was carried out starting with 10 kg of the *meta*-fluorobenzoic acid derivative to provide 3.7 kg of final product **2** with the isolation of only two intermediates (25% yield overall). Although this yield has room for considerable optimization, speed was more important than efficiency at this stage of development.

Prior to 1995, Pfizer was not running nitration reactions in-house due to potential thermal hazard issues. In 1995, a safety group was established with the mandate to define acceptable operating conditions for potentially dangerous reactions, which we used to outsource exclusively. After the appropriate testing was completed, the nitration in Scheme 12 was completed in two 5 kg runs in-house. This reduced the timing for the final bulk delivery by nearly two months compared to sending this reaction to an outside toll manufacturer.

Concerning the reaction itself, the calorimetry was measured for a process where the nitric acid was added to a solution of the starting material in sulfuric acid. This experiment showed that the reaction exotherm was totally dose controlled, essentially a titration, and the heat flow ceased immediately upon the completion of the addition of the nitric acid. The thermal hazard assessment showed that reaction mixtures were unstable above >110 °C. Although the original preparation was being run at 70 °C, RC-1 calorimeter testing showed that the exotherm was totally dose controlled even at 20 °C. Thus, the margin of safety between operating temperature and decomposition onset was increased by 50 °C. The RC-1 was also used to calculate the theoretical amount of ice (48 kg) needed to absorb the adiabatic temperature rise for this bulk run, in case there was an incident and the reaction needed to be quenched immediately. Finally, as an added precaution, the nitric acid was added to the addition funnel in portions.

Scheme 13

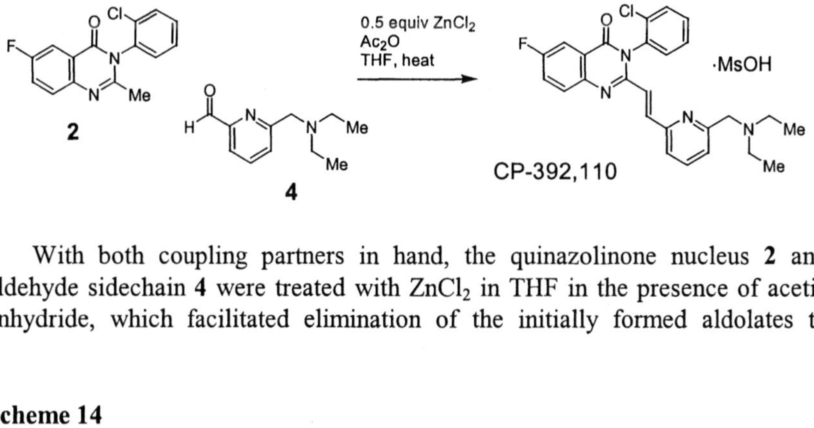

With both coupling partners in hand, the quinazolinone nucleus **2** and aldehyde sidechain **4** were treated with $ZnCl_2$ in THF in the presence of acetic anhydride, which facilitated elimination of the initially formed aldolates to

Scheme 14

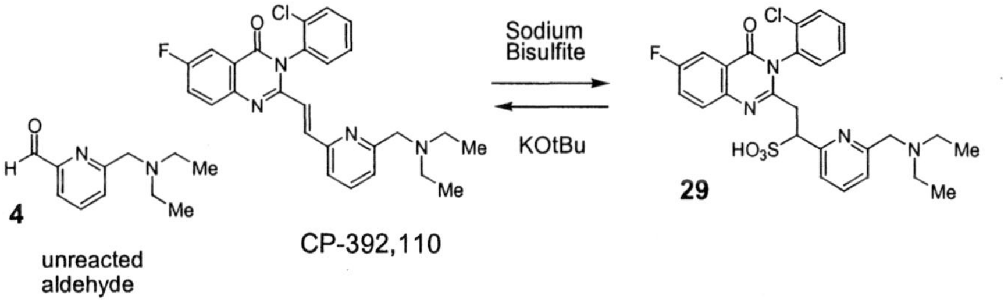

the olefin product (Scheme 13). These aldolates and acylated aldolates, although not isolated and characterized, were observed by HPLC during the course of the reaction. After a fairly extensive screen, it was found that the mesylate salt of CP-392,110 had the best combination of solubility (to enhance bioavailability) and crystallinity (for manufacturability and stability).

After the final coupling to form CP-392,110, we used a bisulfite wash to purge minor amounts of the unreacted aldehyde **4** (Scheme 14). During the course of the addition of the sodium bisulfite, over 50% of the anticipated product CP-392,110 precipitated out as the bisulfite addition complex **29**, as assigned by ^{1}H NMR and mass spec. There is literature precedent for addition of bisulfite across activated olefins, under narrowly defined pH ranges (7). We did not see this side reaction during the pilot, as we were through the optimum pH window (a pH adjustment is part of the work-up) very quickly relative to the longer cycle times during the scale-up itself. This incident highlighted the importance of cycle time considerations when going from pilot to scale-up. In the end, we used this as a purification method and submitted the whole reaction mixture to the bisulfite treatment and isolated all of the desired material as the bisulfite adduct **29**. The complex **29** was then converted back to the olefin CP-392,110 by treatment with potassium t-butoxide in THF.

Scheme 15

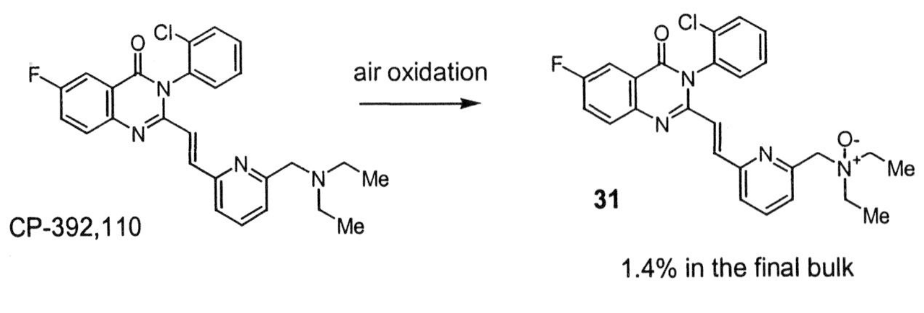

1.4% in the final bulk

Another process related impurity in drug substance was shown to be the N-oxide **31**, as confirmed by ^{1}H NMR and mass spec (Scheme 15). Although not seen in the pilot run, **31** was present at 1.4% in the final bulk lot. Once again, the importance of cycle times and longer exposure to air in going from pilot to scale-up was apparent. Fortunately, as the N-oxide was also a metabolite which formed *in vivo*, this level of the impurity was acceptable.

Returning to the aldol conditions to form CP-392,110 (Scheme 13), we treated the quinazolinone nucleus **2** and the aldehyde **4** with ZnCl$_2$ in THF in the presence of acetic anhydride to facilitate the elimination of the initially formed aldolates. These were the original conditions coming from our Discovery colleagues, and the procedure we used on scale-up. We did, however, spend some time trying to improve this transformation.

Scheme 16

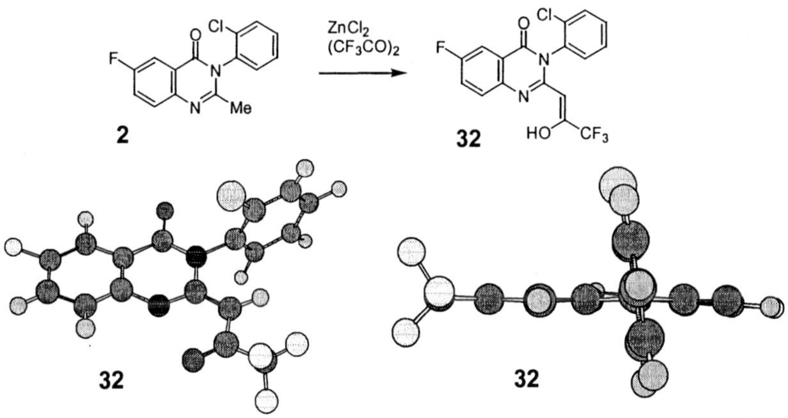

During these optimization studies, an impurity was formed, which was shown to be compound **32** based on ^{1}H NMR and single crystal X-ray analysis (Scheme 16). Compound **32** was the product of acylation of the C_2-methyl

Scheme 17

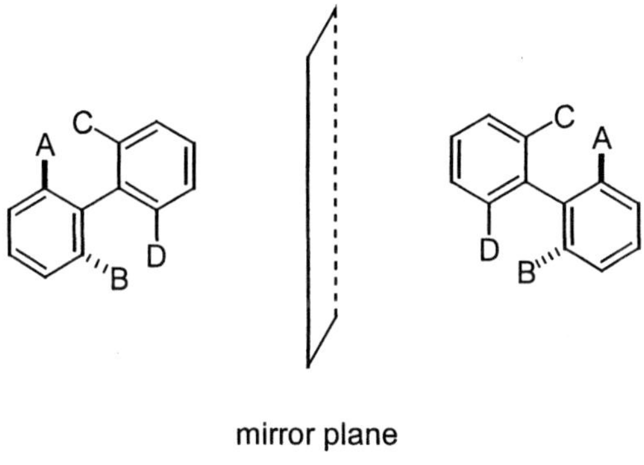

mirror plane

group of the quinazolinone with trifluoroacetic anhydride. What intrigued us most about the X-ray structure was that the edge-on view resembled a typical biaryl bond, which had the the potential for atropisomerism. The question was whether or not the barrier to rotation was sufficiently high to allow for the isolation of the two enantiomeric rotational isomers.

By way of review, atropisomerism is stereoisomerism which exists as the result of hindered rotation about an axis. As illustrated in Scheme 17, if a biaryl bond is unsymmetrically substituted, the rotational isomers exist as non-superimposable mirror images. In this particular case for CP-392,110, rotation about the N_3 bond and the *ortho*-chloro-containing aromatic ring resulted in two enantiomeric rotational isomers. Oki has arbitrarily defined that atropisomerism exists if the half-life of bond rotation is 1000 seconds or greater (*8*). If a rotational barrier is 24 kcal/mol, the half-life is hours at room temp. If the rotational barrier is 31 kcal/mol, the half-life is hundreds of years at room temperature (*9*). The barrier for a typical tribsubstituted biaryl bond, which CP-392,110 closely resembles, is 28 kcal/mol. Hence, we felt that the rotational barrier for CP-392,110 might be high enough to allow us to isolate the atropisomers.

Scheme 18

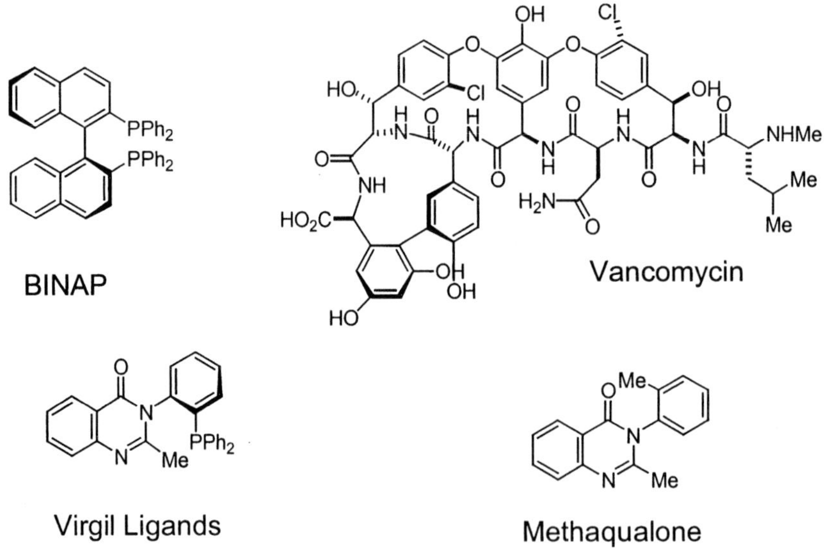

BINAP

Vancomycin

Virgil Ligands

Methaqualone

A number of examples of atropisomerism are apparent from the literature (Scheme 18). BINAP, arguably the most common and general ligand used in asymmtric catalysis, has atropisomerism (hindered rotation) as its basis for chirality (*10*). The antibiotic vancomycin is perhaps the most elegant example of atropisomerism present in nature. In vancomycin there is hindered rotation about both chlorine-containing rings as well as the biaryl bond (*11*). Virgil and coworkers have published their work on the development of ligands for asymmetric catalysis which utilize a quinazolinone nucleus (*12*). Finally, although marketed as a racemate, methaqualone exists as two separable rotational isomers (13, 14).

Scheme 19

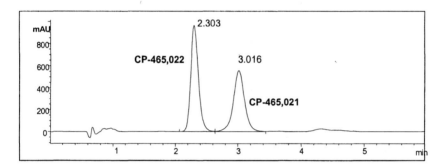

Scheme 20

mirror plane

In order to check for the presence of separable rotational isomers for CP-392,110, we injected a sample onto a chiral stationary phase HPLC column (Scheme 19). Indeed, the compound exists as two enantiomeric rotational isomers with a rotational barrier high enough to allow them to be separated by HPLC.

The next logical experiment was to isolate each enantiomer to see if there was a difference in their biological activities. A semi-preparative separation on a chiral stationary phase was undertaken to obtain 600 mg of each enantiomer. Only one atropisomer, CP-465,022, showed activity in the three tests conducted (1). After preparative separation, we were also able to obtain single crystal X-rays in order to confirm absolute stereochemistry (Scheme 20). The two rotational isomers of CP-392,110 were designated CP-465,022 and CP-465,021. Interestingly, during the course of their SAR studies, our Discovery colleagues made both the des-chloro and di*ortho*-chloro derivatives of CP-392,110, and neither were active. This implied that in the chiral binding site there is room for one but not two chlorines and that one chlorine is essential. Assuming the barrier to interconversion was high enough, these results were also consistent with the biological activity residing in one rotational isomer.

Scheme 21

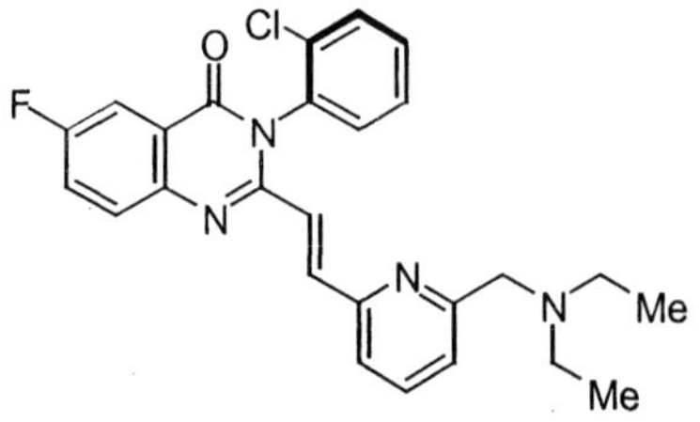

CP-465,022 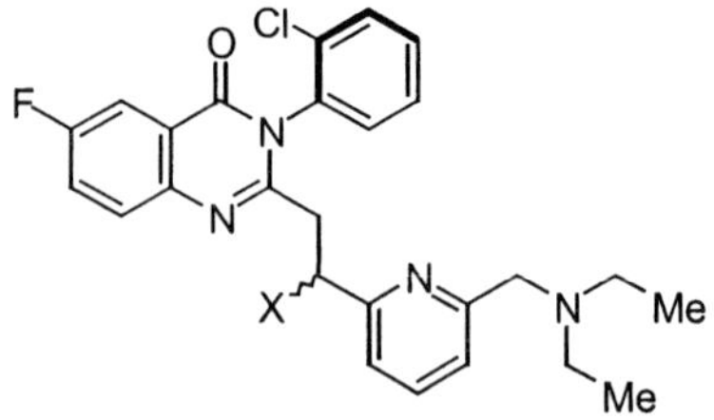   **29a, 29b, 29c,** X = -OH, -OAc, -SO$_3$H

With the benefit of hindsight, were there any other hints along the way that CP-392,110 was actually present as a mixture of enantiomeric rotational isomers with a relatively high barrier to interconversion? During the aldol addition to form CP-392,110 (Scheme 13), we observed peak doubling in the HPLC traces which would be consistent with the diastereomeric aldolates and acylated aldolates **29a** and **29b** (Scheme 21). As the elimination proceeded, both of these sets of peaks disappeared. The same "doubling" was observed by HPLC and NMR for the bisulfite addition complexes **29c**. The chiral center present in these three sets of compounds **29a,b,c** in combination with the axis of chirality produced the diastereomers observed.

Scheme 22

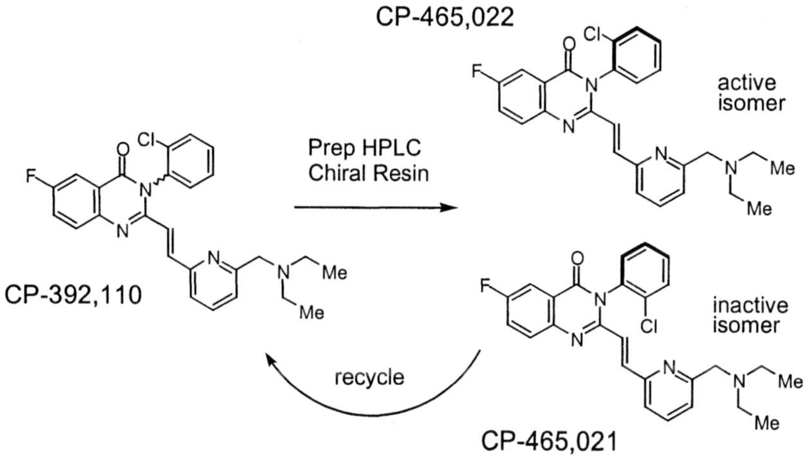

As CP-392,110 was actually a racemic mixture and all of the activity resided in one of the rotational isomers, the candidate nomination was reissued for CP-465,022 to be consistent with the policy of not developing a racemate, unless the enantiomers readily interconvert *in vivo*. We then set about to preparatively separate the atropisomers on a multi-gram scale (Scheme 22). Using a chiral stationary phase, 200 g of racemate was resolved by HPLC to provide 95 g of active enantiomer for further study. We used a 15x20 cm column with a flow rate of 900 mL/minute. The feed was separated in 156 cycles at 20 minutes per run (16.5 mg/mL feed and 80 mL per injection). To speak to the manner in which chiral stationary phases can be utilized in exploratory development for the rapid separation of enantiomers, the total time needed to separate the 95 g of CP-465,022 was less than 72 hours. There was also the potential to recycle the wrong enantiomer, CP-465,021, and one could envision a continuous process where the desired antipode was collected and the undesired one was thermally racemized and added back to the racemic feed.

Although we knew from the preparative separation that the rotational isomers were relatively stable to equilibration, the actual barrier was measured (*9*). The barrier was found to be 28.5 kcal/mol, similar to the expected value for a trisubstituted biaryl bond. This correlates to a half-life of 2.8 years at room temperature. Under physiological conditions (i.e. 37 °C), it would be 171 days, confirming that this compound would have to be developed as the single enantiomer as interconversion under physiological conditions is not rapid. Finally, at 100 °C the half-life is 137 minutes, which is sufficiently short as to allow us to develop a thermal isomerization of the unwanted enantiomer.

Scheme 23

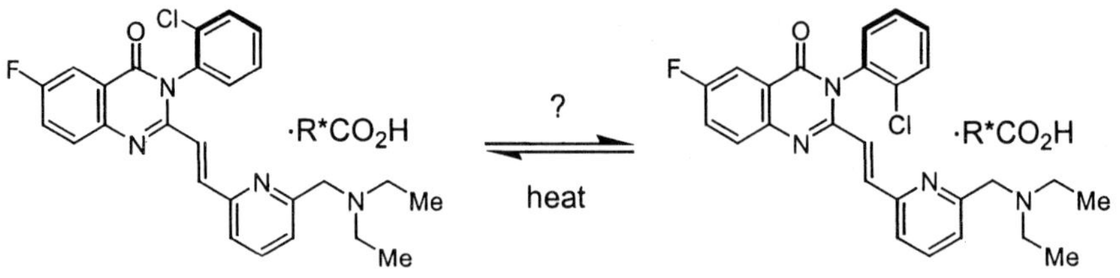

CP-465,022
atropisomer
salt insoluble

CP-465,021
atropisomer
salt soluble

As there was an amine handle for the formation of diastereomeric salts, classical resolution of to obtain CP-465,022 was also an option (Scheme 23). With the ability to thermally racemize the wrong antipode, one could envision a dynamic resolution (or, asymmetric transformation). In the presence of the appropriate chiral acid, there would be an equilibrium of the enantiomers of CP-392,110 in solution, with only the desired antipode CP-465,022 crystallizing out as the insoluble diastereomeric salt. Pfizer has developed a number of other atropisomers, and this sort of dynamic resolution has been worked out for one of those candidates.

While we were in the process of initiating these dynamic resolution studies, we received news that this candidate was going to be discontinued from development due to a lack of efficacy in an *in vivo* model. Although this news is never pleasant to receive, it is the norm rather than the exception at this stage of development. CP-465,022 did suffer attrition, but a number of lessons were apparent:

- We had the opportunity to engage early on technology development and get a convergent route in place for the first regulatory bulk lot. With attrition very likely at this stage, the value of developing a commercially viable route this early is always a topic of discussion. That said, technology development can also impact nearer term timings and costs, and this was clearly the beneficial situation here.

- The impact of longer cycle times and process robustness on scale-up versus the smaller pilots can have a significant impact on impurity generation, as was the case a number of times for this synthesis.
- Although not discussed in detail in this presentation, our Discovery group had theoretical calculations done at the start of this program to determine the rotational barriers in order to see if this series of candidates would exist as separable rotational isomers. The conclusion of those theoretical experiments was that the barrier was too low for the rotational isomers to be separated. Clearly, this conclusion was not borne out in reality.
- The importance of thermal hazard assessment to determine safe operating conditions for hazardous reactions. The ability to do these chemistries in-house can dramatically impact timelines.
- The role of preparative HPLC separation of enantiomers is a major one in exploratory development. Although not always a longer term solution, in the nearer term it is an extremely efficient means to prepare enough single enantiomer to get to a key decision point.

To build on the previous paragraph in terms of lessons learned, there is always a tension in exploratory development with respect to how much longer-term value should be added to a synthetic sequence at the expense of pace and timelines. In the exploratory phase, speed and acceptable quality are the more important concerns, with cost being less of an issue. In the full development phase, pace is less of an issue with quality and economics being the primary drivers. Obviously, if one can add longer term value without impacting timelines, this is desirable in terms of the subsequent bulk campaign(s), and although cost is not a primary driver in the exploratory phase, clearly cheaper is better. As most candidates suffer attrition, however, it is most often not an efficient use of resources to get the "commercial" process in place for the first bulk campaign. Rather, pace to key decision/attrition points is critical, and after some of the early development hurdles have been overcome, the appropriate level of resources are engaged to define a robust commerical process.

This paragraph acknowledges contributions from a broad range of disciplines within Pfizer Central Research. More notably, Thomas Staigers in the Kilo Lab area, who also did a considerable amount of Process Research; our Pilot Plant staff; Brian Vanderplas in Process Research; David am Ende and Pam Clifford in the Hazards Lab for their work on the nitration, which saved the project team considerable time; the Analytical work by Tian Ni to develop the chiral stationary phase separation; measurement of the rotational barriers by Jari Finneman and Lisa Newell; Sam Guhan and Mark Guinn and their labs for the preparative separation of the atropisomers; Jon Bordner and Debra DeCosta in the single crystal X-ray lab for solving a number of structures of impurities;

132

Thomas Sharp and Chandra Prakash for identification of the metabolite; and, finally, to Willard Welch and his team for an excellent and highly productive collaboration with his Discovery laboratory.

References

1. a) Menniti, F. S.; Chenard, B. L.; Collins, M. B.; Ducat, M. F.; Elliott, M. L.; Ewing, F. E.; Huang, J. I.; Kelly, K. A.; Lazzaro, J. T.; Pagnozzi, M. J.; Weeks, J. L.; Welch, W. M.; White, W. F. *Mol. Pharmacol.*, **2000**, *58*, 1310-1317.
   b) Welch, W. M.; DeVries, K. M.; Staigers, T. L.; Ewing, F. E.; Huang, J.; Menniti, F. S.; Pagnozzi, M. J.; Kelly, K.; Seymour, P. A.; Guanowski, V.; Guhan, S.; Guinn, M. R.; Critchett, D.; Lazzaro, J.; Ganong, A. H.; Chenard, B. L. *Bioorg. Med. Chem. Lett.*, **2001**, *11*, in press.
2. Cai, D.; Hughes, D. L.; Verhoeven, T. R. *Tetrahedron Lett.*, **1996**, *37*, 2537-2540.
3. Gilman, H.; Spatz, S. M. *J. Org. Chem.* **1951**, *16*, 1485-1494.
4. Parks, J. E.; Wagner, B. E.; Holm, R. H. *J. Organomet. Chem.* **1973**, *56*, 53, 66.
5. Peterson, M. A.; Mitchell, J. R. *J. Org. Chem.*, **1997**, *62*, 8237-8329.
6. Joshi, K. C.; Singh, V. K. *Ind. J. of Chem.* **1973**, *11*, 430-432.
7. Doering, W. E.; Weil, R. A. N. *J. Am. Chem. Soc.* **1947**, *69*, 2461-2466.
8. Eliel, E. L.; Wilen, S. H.; Mander, L. N. *Stereochemistry of Organic Compounds*; John Wiley & Sons: New York, NY, 1994; Chapter 14.5, pp. 1142-1155.
9. Finneman, J. Pfizer Inc., Groton, CT. Unpublished data, 1999.
10. Noyori, R. *Asymmetric Catalysis in Organic Synthesis*; John Wiley & Sons: New York, NY, 1994; Chapter 1.
11. Evans, D. A.; Wood, M. R.; Trotter, B. W.; Richardson, T. L.; Barrow, J. C.; Katz, J. L. *Angew. Chem. Int. Engl.* **1998**, *37*, 2700-2704.
12. Dai, X.; Wong, A.; Virgil, S. C. *J. Org. Chem.*, **1998**, *63*, 2597-2600.
13. Colebrook, L. D.; Giles, H. G. *Can. J. Chem.*, **1975**, *53*, 3431-3434.
14. Mannschreck, A.; Koller, H.; Stuhler, G; Davies, M. A.; Traber, J. *Eur. J. Med. Chem. Chim. Ther.*, **1984**, *19*, 381-383.

Chapter 9

Practical Convergent Synthesis of SERM 2

Jeffery A. Dodge[1], Kathy K. Ellis[2], Lynne A. Hay[2],
Thomas M. Koenig[2], Charles Lugar[1], Erin E. Strouse[2],
and David Mitchell[2],*

[1]Discovery Chemistry Research and [2]Chemical Process R&D, Lilly
Research Laboratories, A Division of Eli Lilly and Company,
Indianapolis, IN 46285

A convergent synthesis of SERM 2, a selective estrogen receptor modulator is described. This approach incorporates the acylation of 6-methoxy-2-tetralone with [4-[2-(1-piperidinyl)ethoxy]benzoyl]chloride as a key step. In streamlining the process, 6-methoxy-2-tetralone was prepared in situ from 4-methoxyphenyl acetic acid.

In an effort directed at the discovery of SERMs such as trioxifene[1] (2) and raloxifene[2,3] (3) trioxifene and raloxifene, SERM 2 (1) was prepared. This compound exhibits favorable pharmacological properties when compared to the other two compounds.

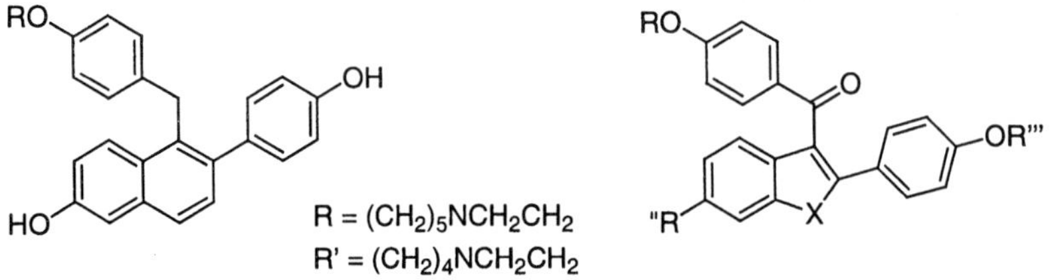

1, SERM 2

2, X = CH_2CH_2, R = R', R" = H
R''' = Me, Trioxifene
3, X = S, R = R, R" = OH
R''' = H, Raloxifene

Figure 1: Slective Estrogen Receptor Modulators (SERM).

A retrosynthesis of **1** is outlined in eq 1. This convergent approach is based on the linear synthesis of **2** and **3**. Compound **2** was prepared from the acylation at C-1 with phenylanisoate followed by conjugate addition of 4-methoxyphenyl magnesium bromide to the enol phosphate.[3] The synthesis was then completed in a stepwise manner including a regioselective demethylation using NaSEt. The synthesis of **3** incorporated the acylation of benzothiophene with the acid chloride **5**. Based on these syntheses, we proposed the acylation of 6-methoxy-2-tetralone with **5**. Addition-elimination of 4-methoxyphenyl magnesium halide, ketone reduction, aromatization followed by demethylation would complete the synthesis.

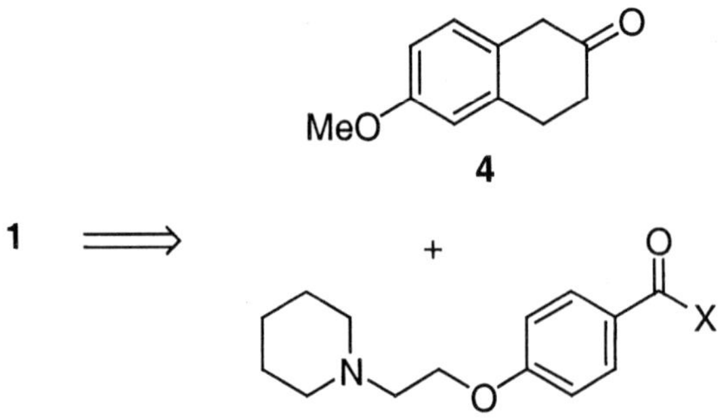

4

5

(eq 1)

Commercially available 6-methoxy-2-tetralone (**4**) did not meet our specification due to the poor quality of material that venders provided. Although pure **4** is a crystalline solid with mp = 33.5-35°C, commercial quantities are available as non-solids with potencies as low as 50%. For our needs, we found the best source of **4** was an *in situ* preparation from (4-methoxyphenyl) acetic acid (**10**), ethylene and AlCl$_3$ in THF.[4] Removal of aluminum salts was accomplished by filtration through a pad of silica gel. The tetralone was then used as a THF solution since the solution was stable at room temperature for more than twenty days. Other conditions such as the use of crude product or a dichloromethane solution were not as stable when compared to the THF solution.

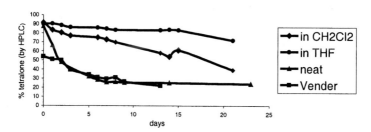

Figure 2: Comparison of 6-methoxy2-tetralone stability in various solvents.

In our system, the acid chloride as the acylating agent was not applicable due to the basic conditions of the proposed transformation and the internal free nitrogen of the piperidyl moiety. However, the trioxifene acylation was adapted by preparing the phenyl ester of **5**. Therefore, the required phenyl 4-[2-(1-piperidinyl)ethoxy]benzoate (**9**) was prepared by alkylating methyl 4-hydroxybenzoate with 2-(1-piperidinyl)ethyl chloride (**6**). Hydrolysis of the methyl ester provided the carboxylic acid that was converted to the acid chloride (**8**) using SOCl$_2$ in a mixture of 1,2-dichloroethane and toluene. Acid chloride **8** was then reacted with sodium phenolate to provide **9** as a stable crystalline solid in 64% overall yield from the carboxylic acid.

Acylation of **4** with phenylbenzoate **9** was performed with sodium hydride in a solution of THF at 0°C, 4 h to provide **11** in 90% yield. No evidence of the isomeric diketone structure or the 1,2-dihydro isomer was observed. The absence of this product may be due to the enol stability of **11** compared to other similar systems where such isomers are observed.

Addition of 4-methoxyphenylmagnesium bromide to the enol ketone **11** was accomplished by first derivatization to the enol phosphate in order to avoid non-regioselective addition to the diketone system; proton abstraction or dehydration of the Grignard adducts. In an attempt to optimize this step, the phosphate was compared to the tosylate, mesylate, methoxy, and benzyloxy leaving groups. A summary of the experimental is presented in the Table I. No product was obtained with the alkyl ethers (Table I, entry c and d). Moderate yields were obtained with the mesylate. The optimum yield was obtained when either the phosphonate or tosylate was used as the leaving group.

Table I: Leaving Groups in the Addition-Elimination Reaction

Entry	Conditions	Yield (%)
a	TsCl, Et$_3$N	68
b	MsCl, Et$_3$N	57
c	MeOH, HCl	0
d	BnOH, HCl	0
e	ClP(OPh)$_2$, Et$_3$N	65

The dihydro naphthalene **12** was aromatized by first in situ reduction of the carbonyl functionality with lithium aluminum hydride in THF at 60°C for 4 h. Treatment of the carbinol solution with HCl gas provided rapid dehydration of the carbinol to the naphthalene ring **13**. The benzo[a]fluorene derivative, cyclization of the 2-aryl ring onto the carbinol group, was not observed.[5] Naphthalene **13** was obtained in 68% yield. Demethylation of the methoxy groups with excess gaseous BCl$_3$ in methylene chloride provided **1** as the HCl salt.

In the pilot plant using 300-gallon equipment, the developed synthesis was executed in five phases without incident. In the first phase, tetralone 4 was prepared and isolated as a THF solution. Next, the second phase, the side chain that was prepared previously, was dissolved in THF and used in the acylation step. The resulting product 11, also isolated as a THF solution, was converted to the phosphate ester before addition of the THF Grignard reagent, the third phase. The fourth phase involved the only isolated solid intermediate. This isolation also provided access to an overall purification by crystallization. Final phase of the campaign was BCl₃ mediated demethylation which provided SERM 2 in greater than 99% purity. Approximately 16 kg of bulk drug was prepared by this approach over a six month timeline that included both laboratory development, engineering and pilot plant processing.

In summary, a highly convergent, practical and efficient synthesis of the selective estrogen receptor modulator SERM 2 is now available that is suitable for pilot plant preparation. Key to the success of the process is the generation of 6-methoxy-2-tetralone. The developed laboratory synthesis is a seven step process with an overall yield of >27%.

Experimental

Compound 4. To a suspension of aluminum chloride (55 g, 406 mmol) in methylene chloride (500 mL) cooled to 0°C, was added 4-methoxyphenyl acetic acid chloride (30 g, 162 mmol) as a methylene chloride solution (250 mL). At that temperature, ethylene gas (13 g, 487 mmol) was introduced into the reaction mixture subsurface over a 10 min period and the reaction allowed to warm to room temperature over a 30 min period. The reaction mixture was then carefully heated to 35-38°C over 30 min and stirred at that temperature for 3 h. Chilled water (5°C, 250 mL) was used to quench the reaction at 0°C and the organic portion separated. The organics were washed with HCl solution (5%, 2X150 mL) water (2X150 mL), EDTA solution (2%, 150 mL) followed by water (150 mL). After drying over anhydrous magnesium sulfate, the solution was concentrated and reconstituted in THF (500 mL). This solution was stored in the freezer for future use.

Compound **11**. Phenyl ester **9** was added as a THF (200 mL) solution to a sodium hydride (25.7 g, 1.07 mmol)/THF (1.6 L) suspension cooled to 0°C. A THF (200 mL) solution of **4** was then added to the reaction mixture at 0°C. After addition of **4**, the resulting reaction mixture was slowly heated to reflux for 2.5 h, cooled to room temperature and quenched with chilled water (1.5 L). The pH was adjusted to 1.0 with HCl (3N, 1.0 L) and the mixture washed with t-butylmethyl ether (MTBE, 1.5 L). The resulting aqueous portion was adjusted to a pH of 8 with sodium hydroxide (5 N) and extracted with MTBE (2X1.1 L). After a brine wash (1.0 L), sodium sulfate was used to dry the combined extracts before concentrating to a residue (85.7 g, 86%).

Compound **12**. Magnesium turnings (5.46 g, 224 mmol) were suspended in THF (25 mL) followed by a crystal of iodine. A THF (100 mL) solution of 4-bromoanisole (30 g 160 mmol) was then added dropwise to the magnesium suspension so as to maintained a modest reflux without external heat. After anisole addition was complete, the reaction mixture was heated to reflux for 2 h and cooled to room temperature.

To a sodium hydride suspension (60% in mineral oil, 1.85 g, 49 mmol) in THF (100 mL) was added **11** as a THF solution (30 mL) at 0°C. After stirring of the reaction mixture for 10 min, diphenylchlorophosphate (10.2 mL, 49 mmol) was added as a THF (10 mL) solution and stirring continued at 0°C for 1 h. The Grignard solution of 4-bromoanisole (92 mL, 0.8 M, 49 mmol) prepared above was added to the reaction at 0°C and stirred for an additional 2 h at 0°C. The reaction was quenched with sulfuric acid (1N, 80 mL) and the pH adjusted to 8-9 with dilute acid or base before extracting with MTBE (300 mL, 2X100 mL). Sulfuric acid solution (1 N, 250 mL) was added to the combined extracts to form the soluble ammonium sulfate salt and the MTBE portion separated. The resulting aqueous solution was adjusted to pH 10 with sodium hydroxide (5 N, 100 mL) and extracted with MTBE (300 mL). After drying with anhydrous magnesium sulfate, the extracts were concentrated to a residue to provide **12** as a colorless oil.

Compound **1**. Boron trichloride was condensed (33 mL) using a cold finger apparatus and added in one portion to a suspension of **12** (15 g, 29 mmol) and 1,2-dichloroethane (170 mL) cooled at 0°C. The resulting purple mixture was sealed with a stopper and allowed to stir for 14 h under the pressure generated by BCl$_3$ at room temperature. At 0°C, the reaction was quenched by dropwise addition of methanol added through a condenser (caution, vigorous reaction). The product precipitated upon methanol addition and was collected by filtration to provide 12.3 g (86.6 %) of **1** as the hydrochloride salt, a white crystalline solid.

References

1) Jones, C. D.; Suarez, T.; Massey, E. H.; Black, L. J.; Tinsley, F. C. *J Med. Chem.* **1979**, *22*, 8, 962.

2) Jones, C. D.; Jevnikar, M. G.; Pike, J. A.; Peters, M. K.; Black, L. J.; Thompson, A. R.; Falcone, J. F.; Clemens, J. A. *J. Med. Chem*, **1984**, *27*, 8, 1057.

3) Jones, C. D.; Blaszczak, L. C.; Goettel, M. E.; Suarez, T.; Crowell, T. A.; Mabry, T. E.; Ruenitz, P. C.; Srivatsan, V. *J. Med. Chem.* **1992**, *35*, 5, 931.

4) Sims, J. J.; Selman, L. H.; Cadogan, M. *Org. Synth.* Coll. Vol. 6, 744 (1988).

5) Durani, N.; Jain, R.; Saeed, A.; Dikshit, D. K.; Durani, S.; Kapil, R. S. *J. Med. Chem.* **1989**, *32*, 8, 1700.

INDEXES

Author Index

Subject Index

W

Highlights from ACS Books